Understanding Metocean

Understanding Metocean

A User Guide for Offshore Renewables and Oil & Gas

Ian M. Leggett

BSc, CSci, CMarSci, CMarTech, FIMarEST
Managing Director, OceanExpert Limited

WHITTLES PUBLISHING

Published by
Whittles Publishing,
Dunbeath,
Caithness KW6 6EG,
Scotland, UK
www.whittlespublishing.com

ISBN 978-184995-499-0

Printed and bound by CPI Group (UK) Ltd, Croydon, CR0 4Y

Contents

Acknowledgements

I would like to take this opportunity to acknowledge and thank some of the many colleagues and friends who I have worked with including.

- Geoff Haskins who gave me my first job at Shell in 1978.
- Chris Graham who introduced me to Metocean and taught me so much.
- Jon Upton, Graham Feld and Rizwan Sheikh who I recruited in Aberdeen in the 1990's – we were such a great Metocean team. Additional thanks to Jon for providing input to this book and reviewing drafts.
- Rupert Hunt who always valued my input and expertise in Aberdeen.
- Colin Grant for being a long-standing colleague and friend, and providing initial thoughts and comments on this book.
- The many colleagues from the Metocean contractor community including Ian Bellamy, Nigel Bellamy, Frank Beiboer, Joe Fox, John Mitchell, Peter Owrid, Ralph Rayner, Robin Stephens, Ian Wade and Andrew Watson.

I would also like to thank SSE plc for supporting me in producing a Metocean Procedures Guide (MPG) for them in 2012 that sowed the seeds for this book.

Preface

Having spent a lifetime in the metocean business I thought it was about time that I put pen to paper and share some of my experiences and expertise to help shine a light on what it's all about. Hopefully this book will provide useful information not only for those new to the offshore business but also for those who have been round the block more than once.

It has taken longer than I thought. Ideas began to crystallise in 2014 but work and family commitments, plus Covid and a number of re-writes and updates, have slowed things down. However, finally, here it is: everything you need to know about metocean but were afraid to ask! Well, maybe not everything – that would require a great many more pages.

What is metocean?

For those unfamiliar with the term, metocean is derived from *met*eorology (the study of the weather) and *ocean*ography (the study of the sea). It is both an activity and a discipline and is concerned with quantifying the impact and effect of the weather and the sea on people, activities and infrastructure relating to the offshore renewables and the oil and gas industries. (It should be noted that for the purposes of this book 'infrastructure' is a generic term referring to any man-made object installed and operating offshore: for example, wind turbines, fixed platforms, rigs, floating installations, subsea installations, pipelines etc.)

Although the term metocean may appear wide ranging, it is important to note that it is specifically concerned only with the physical attributes of the atmosphere and the sea (as opposed to the biological and chemical attributes). The major physical attributes include the following parameters:

- wind speed and direction
- wave height, period and direction
- current speed and direction
- water level variations

There is also a wide range of other parameters, including air and sea temperature, snow and ice, and visibility.

Quantifying the impact of these physical attributes can be facilitated by following a standard metocean process or procedure, which comprises three key phases:

- an *operational* phase, concerned with data acquisition and weather forecasts
- a *data management* phase, concerned with quality control and the archiving of data
- an *analytical* phase, concerned with the provision of statistical reports, studies, information and advice from the data

Book structure

Although both offshore renewables and the oil and gas industries have their own approaches and requirements for offshore developments and operations – be that for a wind turbine or an oil-producing platform – the metocean needs and requirements are very similar.

This book expands on the three key phases of the metocean process or procedure mentioned above, and supplements this with detailed information on metocean terminology, definitions and other related topics, to show how these needs and requirements can be met via deliverables associated with the lifecycle of offshore projects.

It should be noted that this book is not intended to be a Metocean standard or policy document, although reference to appropriate standards, guidelines and policies is made as and when applicable. Rather, it is intended to be a source of information to give people a better understanding of what metocean is all about and how optimum use of data and information can benefit their activities. It should therefore complement existing Metocean standards.

As the discipline of metocean covers a wide range of topics and activities, it is also not intended that this book cover every aspect in detail. Rather, it seeks to address as many of these as possible and highlight the key aspects.

Although the book is written primarily from a UK perspective, much of the material is equally applicable in other parts of the world.

Readership

Although metocean appears to be a unique activity, the data and information generated are required by a diverse range of people. The intended readership is therefore wide and the book is aimed at both metocean providers – responsible for producing data and information – and end-users who require a range of products and services.

'Providers' includes metocean staff employed by offshore renewables (OR) and oil and gas (O&G) companies, together with those working for contractors, government and academia.

'End-users' cover a wide range of disciplines and activities, including design, structural, pipeline, subsea and well engineers, project managers, operations supervisors, environmental scientists, surveyors and logistics. Students, graduates and new entrants to OR and O&G may also find the book useful.

Note

Throughout the book there is a variety of terms such as Hs, Hmax, Tp, Tz, σθ etc. where, in an academic context, the variable would be italicised and the remainder of the term would be set as a subscript, thus: H_{max}. In the practical and applied aspects of metocean, the general practice is to dispense with these refinements and so, this book being a practical introduction the non-italicised style has generally been adopted.

To my wonderful wife Yvonne, who has supported me throughout this endeavour

Chapter 1
An introduction to metocean

This chapter helps to set the scene and introduce the reader to metocean. It contains six main sections:

- a brief history of metocean
- why metocean is needed
- a summary of the key metocean activities
- an integrated approach to metocean activities
- metocean expertise
- standards, policies and guidelines

A brief history of metocean

As the oil and gas (O&G) industry expanded its area of interest from onshore to offshore in the 1960s, a key challenge was posed in the requirement to design and operate installations in a potentially hostile marine environment. Amongst other things, this meant that a wide range of data and information was required on historical, present and future weather and sea conditions.

There was no expertise in these subjects within the O&G industry at this time; instead there was reliance on governmental organisations and a very limited commercial sector to obtain such information and assistance.

As the industry expanded further in the 1970s, the requirements for information on sea and weather conditions grew. This resulted in some of the larger O&G companies reviewing whether they should establish their own in-house expertise. Shell was the first company to do so, setting up a department in The Hague in the Netherlands in the mid-1970s to co-ordinate these activities globally. This department was called Metocean Services to provide a shortened, more user-friendly form of meteorology and oceanography.

Other major O&G companies soon followed Shell's example and during the 1980s metocean departments were established in most of the larger companies, including Amoco, BP, Chevron, Conoco, Elf, Exxon, Marathon, Mobil, Norsk Hydro, Statoil, Texaco and Total.

The expansion of metocean expertise and requirements within the O&G companies led to a commensurate expansion in the availability of metocean contractors and consultants to meet their needs.

Since 2000, there has been significant consolidation in the O&G sector, plus the emergence of many newer smaller operators. Whereas the larger companies still retain internal metocean expertise, it should be noted that most other companies have not established such in-house capabilities, relying instead on external metocean contractors and consultants.

Another important development since 2000 has been the move to so-called 'frontier' environments: namely, deep-water (500–3000 m) and polar/cold regions. Whereas metocean expertise is readily transferrable to the deep-water situation, additional expertise is often needed for polar/cold regions to deal with specific ice issues.

Also, over the last 20 years metocean has expanded into the offshore renewables (OR) sector as it has similar requirements for data and information. OR companies are benefitting from the experience built up in the O&G industry over the last 40+ years, and some now have their own in-house metocean staff.

Why metocean is needed

All offshore activities, operations, and infrastructure such as wind turbines and oil-producing platforms, are impacted in varying degrees by the weather and sea conditions. The variability of such conditions also poses many challenges for those working offshore. For example, whilst it is well appreciated that storm conditions can adversely affect activities, calmer conditions can also be a concern if accompanied by poor visibility.

This means that the availability of continuous high-quality real-time data, reliable weather forecasts and accurate metocean statistics can play a major role in supporting the key business drivers of reducing costs and improving safety by providing a range of appropriate deliverables. Some examples of these are:

- real-time data of wave heights and wind speeds can assist with operational decision-making: for example, whether a vessel can offload cargo or maintain a wind turbine
- reliable weather forecasts can help to plan operations up to one week ahead to minimise downtime
- operational statistics such as monthly weather downtime can help in planning and costing future activities several months or years ahead: for example, rig selection, installation scheduling and O&M phasing
- accurate metocean design criteria for 50-year or 100-year return periods can optimise the design and contribute to the structural integrity of a wide range of offshore infrastructure from wind farm layout through to individual turbines as well as O&G installations

These and other deliverables are discussed in more detail in Chapter 3. Failure to provide such deliverables in a timely fashion and with accurate inputs/outputs can have serious safety and cost implications. In this respect two further points may be made:

- There is a need to strike the right balance between safety and cost, where appropriate. For example, an over-estimation of design criteria could result in increased costs for a platform or wind turbine, whereas under-estimation could have consequences for safety and integrity.

- Investment in metocean at an early stage of a project or activity will result in significant cost savings in later phases. It is not always easy to quantify exact cost savings from optimum use of metocean data and information, but the lifetime savings could run into millions of pounds.

Finally, it should be appreciated that, apart from cost and safety issues, the provision of metocean deliverables is also required for regulatory compliance in many parts of the world.

Key metocean activities

The Preface introduced the concept of a standard metocean process or procedure consisting of three key phases. However, in terms of deliverables these phases can be expanded to six activities, which form the basis of the requirements for the end-user deliverables (discussed in more detail in Chapter 3):

- Activity 1: acquiring data
  - that is already available
  - through measurement by metocean sensors deployed offshore
  - by developing numerical models to simulate past conditions
- Activity 2: providing real-time display and transmission of data from offshore measurements
- Activity 3: managing the provision of weather forecast services
- Activity 4: performing quality control checks on the data, archiving it into a database and ensuring their long-term management and availability
- Activity 5: carrying out studies and statistical analyses of the archived data to provide two main types of output
  - operational statistics, focusing on typical conditions
  - design criteria, focusing on extreme conditions
- Activity 6: providing briefing notes and expert advice

These activities tend to follow sequentially from a timing point of view. However, activities 3 (weather forecasts) and 6 (briefing notes and expert advice) can occur at any time.

There are also other activities or subjects directly or indirectly related to metocean, including environmental issues, sediment transport, marine growth, acoustics and glaciology. They are usually addressed by experts in the specific fields, such as environmental or geotechnical people.

Some specific elements of glaciology (particularly at the boundaries of sea and ice) are directly associated with metocean for the O&G industry.

An integrated approach to metocean activities

It is important to stress that there is interdependence amongst many of the metocean activities and deliverables. For example, accurate operational statistics and design criteria require good quality, long-term data. (What is meant by 'long-term' is discussed in Chapter 4, Metocean Data Acquisition). Similarly, the accuracy of weather forecasts can be improved by ensuring real-time data are sent to forecast providers quickly.

The three key phases already outlined should therefore be seen as a continuous activity akin to a business process, and viewed as an integrated metocean approach. This can be summarised as follows:

Metocean Business Process
(operations → data management → analyses)

An integrated approach (Figure 1.1) makes sense, particularly when considering different end-users of metocean information. For example, logistics and O&M personnel are primarily interested in real-time data for vessel, helicopter and maintenance operations, whereas project engineers are more interested in extreme 50-year or 100-year values for the design and reassessment of O&G installations or wind turbines.

In order to satisfy such differing requirements it is vital that real-time data are also recorded, quality controlled and archived for future use. Thus, when thinking about metocean requirements specific to a particular end-user, it is important that consideration is given to other potential end-users as well.

Other issues to consider with respect to an integrated approach include:

- ensuring data collection starts early in the offshore development lifecycle so that there is sufficient time to acquire enough data
- selecting technically competent metocean contractors to carry out specific activities
- encouraging the contractors to work together where appropriate to maximise data returns, ensure data quality and improve forecasts

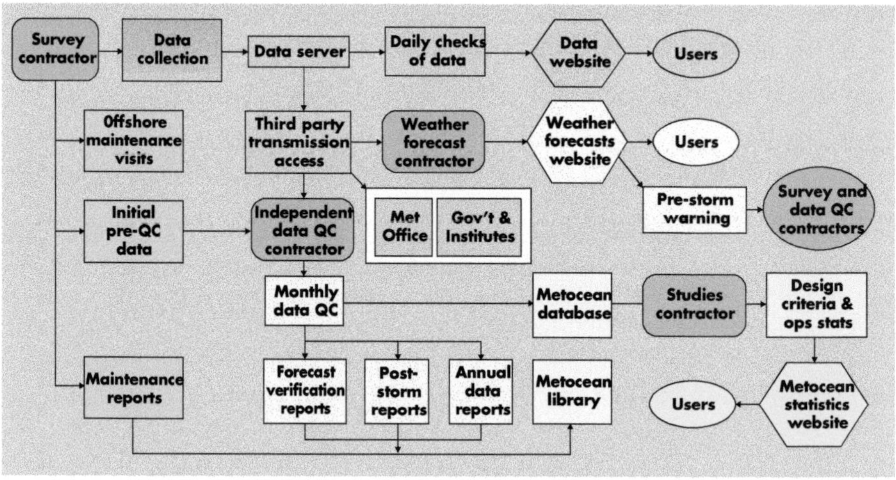

Figure 1.1 An example of an interconnected, integrated approach to metocean activities. It shows the services, products, linkages and co-operation between four contractors (Survey contractor, Weather forecast contractor, Independent data QC contractor, Studies contractor).

Some of the benefits of these linkages include:

- the survey contractor responsible for data collection ensures that data are also made available to weather forecast providers to improve the accuracy of forecasts
- the weather forecast contractor alerts the survey contractor to any upcoming bad weather for planning offshore maintenance visits
- the independent quality control (QC) contractor ensures a quality dataset is available for the studies contractor

Metocean expertise

OR/O&G companies

Metocean is a niche discipline and, in this respect, the number of practitioners employed in OR/O&G companies globally is relatively small. A few companies do have internal expertise as discussed previously, but many do not, tending to rely on external expertise from metocean contractors and consultancies. For the latter situation it may be prudent to consider appointing a metocean focal point, either within the company or by using an external specialist to provide the necessary co-ordination and direction.

Metocean contractors and metocean consultancies

There are different types of providers of commercial metocean services. These range from individual consultants, generally offering specific support on particular topics or subjects, to companies that provide services to collect, archive and analyse data. Although there is no hard and fast rule, the former tend to be referred to as metocean consultancies whilst the latter are referred to as metocean contractors.

The activities of metocean consultancies tend to be mostly associated with:

- specialised data analyses and desk studies, particularly with respect to design criteria
- independent reviews of metocean contractors' output

The activities of metocean contractors fall into one or more of the following:

- site-specific measurements
- numerical hindcast modelling
- weather forecasts
- operational statistics and design criteria (desk studies)

Some contractors offer a comprehensive service covering all categories, whereas others occupy specific niches: for example, some only carry out site-specific measurements whilst others only undertake desk studies.

Although it may appear beneficial from a contract perspective to utilise one contractor for all metocean requirements, the contractor may not necessarily be a top-quartile performer for each activity. It is therefore prudent to consider using different contractors for different activities, as per the example for the integrated approach in Figure 1.1.

Thus, when selecting potential contractors (or consultants) due account should be taken of their expertise and experience. In addition, it is vital to consider contractors HSE issues and record, particularly with respect to offshore work.

It should be noted that, aside from commercial consultants and contractors, there are also a number of institutes, governmental and quasi-governmental organisations with metocean expertise.

Metocean standards, policies and guidelines

There are a number of standards, policy and guideline documents issued to inform the OR and O&G sectors of the importance of appropriate metocean input to assist with optimising design and operational requirements. These documents have been issued by organisations such as:

- American Petroleum Institute (API)
- UK Civil Aviation Authority (CAA)
- Det Norske Veritas (DNV)
- UK Health and Safety Executive (HSE)
- Helicopter companies
- International Electrotechnical Commission (IEC)
- International Association of Oil and Gas Producers (IOGP)
- International Organization for Standardization (ISO)
- Norsk Sokkels Konkuranseposisjon (NORSOK Norwegian O&G standards)

Individual OR and O&G companies may also have their own additional standards, policy and guidelines documents.

As there are a wide range of different documents it is recommended that guidance is taken from a metocean specialist as to which are best to apply. This is because misinterpretation or misuse of the information could lead, for example, to non-compliance or over-conservatism in design. In most cases, however, ISO, DNV and IEC standards will take precedence. A comprehensive list of these documents is provided in Appendix 1.

It should also be noted that such documents may be updated periodically, or even withdrawn, so users should always ensure that they are referencing the most up-to-date or appropriate versions.

Chapter 2
Metocean terminology

It was noted earlier that metocean is concerned with the physical attributes of the atmosphere and ocean, such as winds, waves and currents. It is important at this point to explain how these attributes and their output can be categorised. For the purposes of this book there are five main ways:

- *data type*: another way of describing the physical attributes of the atmosphere and ocean – for example, winds, waves, currents and water levels
- *parameter*: a way of quantifying elements or descriptors of the data types – for example, significant wave height and mean wind speed
- *data*: individual measurements or model outputs of parameters at discrete time-steps, such as hourly
- *dataset*: a record of measurements or model outputs of parameters at discrete time-steps for a given time period – for example, 5 years of measured significant wave heights and mean wind speeds
- *information*: visual displays of processed data or statistical outputs and reports from analysed data

The collection of data to generate datasets is addressed in more detail in Chapter 4. This current chapter focuses on the key parameters associated with the four main data types, i.e. winds, waves, currents and water levels. It provides explanations of these parameters and, where appropriate, gives additional descriptive information to enhance understanding of various processes and related phenomena.

It is important to realise that there are a considerable number of different parameters associated with each of these data types, and that clarity of terminology is vital to avoid any misunderstandings in the request for, and use of, metocean information.

Other metocean terms and topics that require explanation are provided in Appendix 2. Appendix 2 also includes terminology and descriptive information relating to the nearshore zone, where shallowing water, complex bathymetry and coastal configuration have major impacts on the main data types and a number of different processes and phenomena can occur accordingly.

Data types and main parameters

The key parameters for the main data types are:

Winds

- mean wind speed and gust wind speed for various durations between 1 hour and 3 seconds, and various elevations above sea level (ASL)

Waves
- significant and maximum wave height
- wave crest elevation
- mean, peak and associated wave period
- mean and peak wave direction

Currents
- mean current speed and direction profiles

Water levels
- tidal level
- surge level

Some of the individual parameters can be combined to make additional parameters. For example:

- still water level (SWL) – combining tide and surge
- extreme water level (EWL) – combining tide, surge and wave crest

There are also two collective data types – 'other meteorology' and 'other oceanography' – which encompass the remaining metocean parameters:

Other meteorology
- air temperature; barometric pressure; cloud cover and height; humidity; icing; lightning; precipitation (rainfall and snow); visibility

Other oceanography
- salinity; sea temperature; sea ice

1. Winds

Mean wind speed

The mean wind speed is a sustained wind speed measured over a given period of time at a given height above the sea surface. It will also have an associated direction. There are three main parameters:

- 1-hour mean: the mean wind speed measured over 1 hour
- 10-minute mean: the mean wind speed measured over 10 minutes
- 1-minute mean: the mean wind speed measured over 1 minute

As winds are normally sampled at a rate of 1 Hertz (1 Hz), i.e. 1 sample per second (see Chapter 4), a 10-minute mean wind speed would be derived from an average of 600 1-second samples.

The 10-minute mean is also referred to as U10 and is generally used as the standard sampling period. Note that 10-minute means can refer to one sample of 10 minutes each hour or to six separate 10-minute means every hour.

Gust wind speed

The gust wind speed is the highest wind speed within the mean wind speed sample duration averaged over a number of seconds. The main parameter is:

- 3-second gust: the highest speed in a sampling duration, averaged over three consecutive 1-second samples

2. Waves

Before explaining the main wave parameters, it is useful to first consider four aspects of waves:

- types of wave
- wave characteristics
- sea state
- wave spectrum
- steepness

Types of wave

Although most people think of waves as those generated by wind, there is in fact a wide range of waves, which can be classified according to their frequency or wave period and the generating mechanism. These can range from capillary waves or ripples with wavelengths of a few centimetres, to earthquake-generated tsunami with wavelengths in excess of 100 km.

It should also be noted that waves are not just a surface feature: they can also exist at depth in the ocean. Such waves are known as internal waves.

This book focuses on surface waves generated by the wind, as these are the types of wave of primary interest to OR and O&G companies. However, further information on tsunami and internal waves is provided in Appendix 2.

Wave characteristics

Sea surface waves are caused by wind, and can be quantified in terms of height, crest, trough, wavelength, speed, period and direction (Figure 2.1).

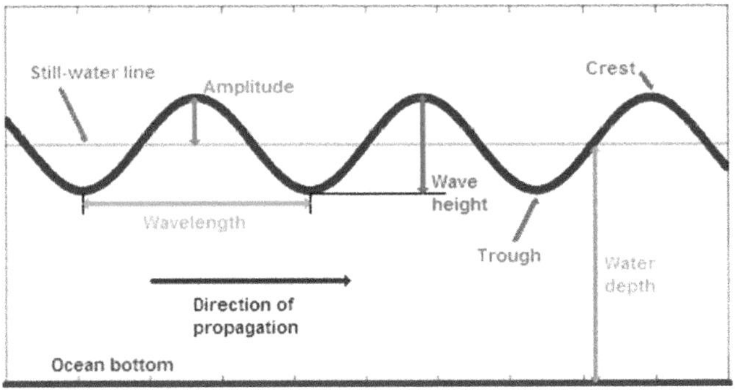

Figure 2.1 Wave characteristics.

Wave amplitude

The amplitude of a wave is the maximum height of the wave crest above the level of 'calm water' or the maximum depth of the wave trough below the same level. The wave height is effectively twice the amplitude.

Wavelength

The wavelength (L) is the distance from any point on a wave to the same point on the next wave. Wavelength = $1.56T^2$, where T is the wave period.

Wave speed

The wave speed is how far the wave travels in a certain time and is expressed in metres per second (m/s). It is a function of the frequency (wave period T) and the wavelength (L). Clearly, the speed $c = L/T$.

Wave steepness

Wave steepness is a characteristic of individual waves and is calculated as the wave height divided by the wavelength. Steepness is typically between 1:10 and 1:30. Note that waves tend to break if the steepness is <1:7.

Wave height, period and direction are explained in more detail later in this chapter.

Sea state

Sea state is a term used to describe the condition of the sea during a period of time in which its statistics remain approximately stationary or unchanged. The time period is often assumed to be 3 hours, although it depends on the particular weather situation at any given time.

The sea state can be broadly specified by two components: swell and wind-sea.

Swell

Swell is the component of the sea state related to waves generated by winds remote from the site and having travelled to the site.

Wind–sea

Wind–sea is thc component of the sea state related to locally generated winds.

Wave spectrum

The wave spectrum essentially defines the distribution of wave energy among different wave frequencies (equivalent to wave periods).

Most wave parameters can be derived from the analysis of spectra. For example, the area under the wave spectrum is the zeroth spectral moment $m0$, which is a measure of the total energy in the sea state; $m0$ is also used in contemporary definitions of significant wave height (Hs).

The wave frequency spectrum (integrated over all directions) is often described by the use of some parametric form such as the Pierson–Moskowitz spectrum or the JONSWAP wave spectrum.

Spectral moment

An integral over frequency of the spectral density function multiplied by the *n*th power of the frequency, either expressed in Hertz (Hz, cycles per second) as

$$m_n(f) = \int_0^\infty f^n S(f)df$$

or expressed in circular frequency (radians/second) as

$$m_n(\omega) = \int_0^\infty \omega^n S(\omega)d\omega$$

Further details on wave spectra can be found in Appendix 3.

Wave parameters – height

Significant wave height (Hs or H*m*0)

Hs is the standard measure of wave height used for operational decision-making and in the derivation of operational statistics and design criteria.

It was originally defined as the mean height of the highest one-third of the 'zero up-crossing' waves in a sea state (H1/3). Today it is more accurately described as a statistical measure of the height of waves typically sampled over 17 or 34 minutes, derived from spectral analysis, as 4*√*m*0 (H*m*0), where *m*0 is the zeroth spectral moment.

Thus, although significant wave heights are presented as absolute values, it should be appreciated that such values are invariably statistical representations of reality (Figure 2.2).

This means, for example, that if we have a calculated Hs of 3.5 m from a 17-minute sample, the range of heights for all the individual waves during the sampling period (which is usually between 100 and 200 waves) will be anywhere between, say, 0.5 m and 7 m.

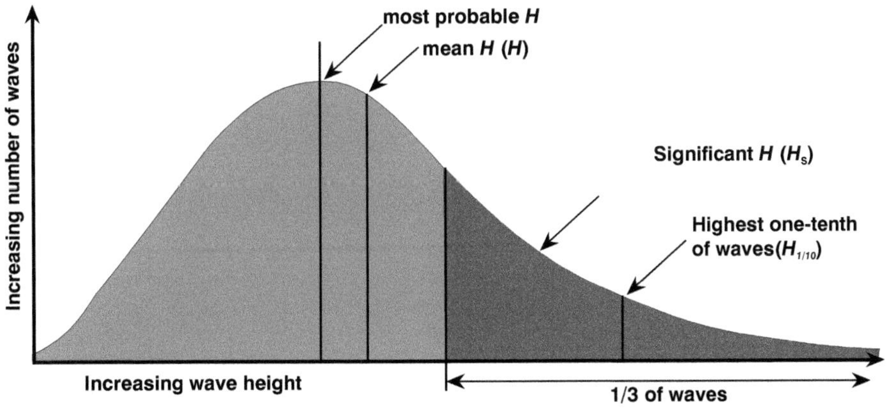

Figure 2.2 Statistical wave distribution.

Hs is therefore a 'synthesis' of all the individual waves, and is approximately equivalent to the 87%tile of those individual waves.

Maximum wave height (Hmax)

Hmax is nominally the largest individual crest-to-trough height within a measured sample (otherwise known as Hmax_T). It can also be expressed statistically as the highest wave in a sample assuming a particular distribution of individual waves – for example a Rayleigh distribution.

It should be noted that the Hmax used in design is not necessarily the highest wave you could ever expect: for example, the 100-year return period Hmax is normally expressed as the 'most probable' Hmax (3 hours).

The relationship between Hs and Hmax is not fixed, although typically Hmax is from 1.5 to 2.5 times Hs, depending on the sea conditions. This relationship is explored further in Chapter 7.

Maximum crest elevation (Hc or Cmax)

Cmax is the maximum upward displacement of a wave above a datum (for example, mean sea level, MSL). Note that for most wave conditions crest/trough asymmetry is evident. Thus a typical wave height is made up of approximately 60% crest and 40% trough.

Note that wave crest elevation is also a parameter used in calculations of water level (see Section 4).

Wave parameters – period

Wave period is the time interval between the arrival of consecutive crests at a stationary point. It is notated as T, and there are three main parameters: average or mean zero-crossing period; spectral peak period; associated wave period.

Average or mean zero-crossing period (Tz)

Tz is the average period between waves crossing a mean sea level line (effectively 'zero') in either an upward or downward fashion in a sea state. In this respect the more usual approach is assessing upward and is referred to as zero up-crossing period is often estimated from the zeroth and second moments of the wave spectrum using the following equation:

$$T_Z = T_2 = \sqrt{m_0(f)/m_2(f)} = 2\pi\sqrt{m_0(\omega)/m_2(\omega)}$$

Spectral peak period (Tp)

Tp is the period of the maximum (peak) energy density in the wave spectrum. It is typically 20–40% higher than Tz.

Associated wave period (Tass or Tmax)

Tass or Tmax is a period parameter that was traditionally used or 'associated' with Hmax when wave parameters were derived in the time domain as opposed to the frequency domain. It was obtained by applying ratios to Tz. Since the use of spectral methods in the frequency domain this parameter has tended to be superceded by Tp. In terms of how it compares to Tp, Tmax is typically about 0.9Tp.

Wave parameters – direction

Wave direction is the direction from which waves come. It can be specified by two main parameters: mean and peak wave directions. It can also be expressed in terms of how the energy is spread directionally.

Mean wave direction

Mean wave direction is the mean of all the individual wave directions in a time-series representing a certain sea state.

Peak wave direction

Peak wave direction is the wave direction associated with the peak energy of the wave spectrum.

Wave spreading

The directional spreading function $D(\theta)$, describes the directional distribution of wave energy in irregular multi-directional sea states. It can be quantified in terms of the principal direction of wave propagation θp, and the directional spread or standard deviation of the spreading function $\sigma\theta$.

3. Currents

Before explaining the main current parameters it is useful to set the scene by considering briefly the forcing mechanisms that cause/generate currents.

Currents are intrinsically linked to water levels. Thinking in terms of simple mass conservation, in order for water levels to rise or fall the water has to flow into or out of an area. Current is therefore a generic term that describes this net movement of water. The movement is brought about by the combined effects of:

- astronomical forcing (the Sun and Moon)
- meteorological forcing (caused primarily by wind)
- oceanographic forcing (caused primarily by density differences)

The strength, impact and presence of these forcing mechanisms vary on a global basis temporally, spatially and vertically. Subsequent to these initialising forces the current flow will then be modified by local and regional conditions, such as water depth, coastline configuration, and friction. Further details on the forcing mechanisms can be found in Appendix 2.

Main current speed parameters

The forcing mechanisms result in currents consisting of two key components:

- a tidal component (due to astronomical forcing)
- a non-tidal or residual component (due to the other forcing mechanisms)

Tidal current

The tidal current is that part of the total current (see below) related to the astronomical forcing of the Sun and the Moon. Tidal currents are predictable and are also referred to as tidal streams. Speeds can vary from less than 0.2 m/s in the open ocean to more than 3 m/s in some nearshore waters.

Residual (non-tidal) current

The residual current is that part of the total current that is not due to tidal forcing. Residual currents are caused by a variety of physical mechanisms. They are comprised of a large range of natural frequencies and magnitudes in different parts of the world, and are generally not predictable.

Some residual currents are due to meteorological forcing (primarily the wind) whereas others are due to oceanographic forcing. Examples are provided in Appendix 2.

Residual currents are also sometimes referred to as surge currents, particularly where meteorological forcing dominates.

It is important to note that residual or surge currents should not be confused with the phrase 'long-term residual drift', which refers to the equivalent speed an object in the sea moves at from point *a* to point *b* over a finite time period. In reality it is not a true speed as the object may meander and not travel in a straight line.

Total current

The combination of the tidal and residual components is referred to as the total current. This is what is observed and measured at sea, and is the parameter provided in operational statistics and design criteria reports.

The total current as measured will typically consist of 10-minute mean values but other sampling regimes can be used depending on the location of interest and any processes that may need to be identified: for example, high-frequency events may need 1-minute samples.

Summary of current components

Figure 2.3 is a much simplified overview of the components of current flow and how they make up the total current.

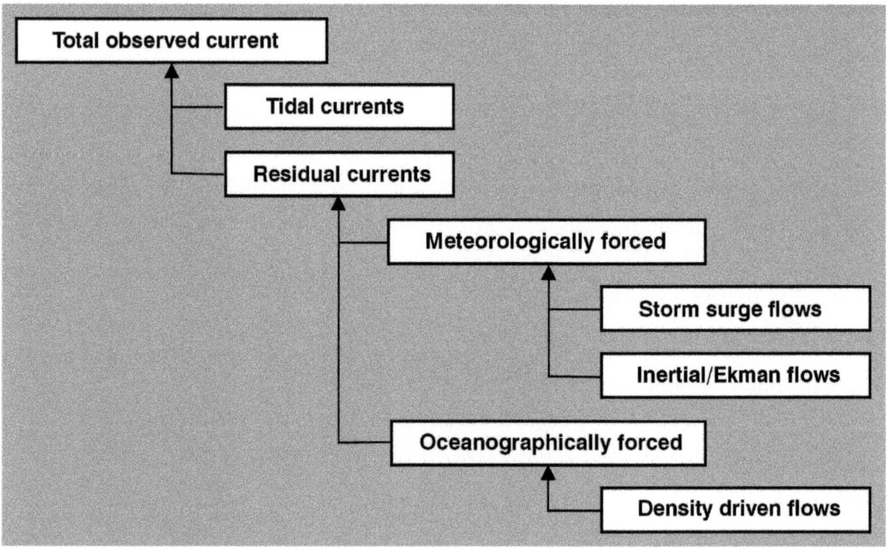

Figure 2.3 The components of current flow.

4. Water levels

As the seas and oceans are in constant motion, the sea level varies on a wide range of timescales. An average level of the sea surface is known as mean sea level (MSL) and this is used as a reference vertical datum. It is essentially the midpoint between the mean low and high tides.

Variations in sea level are principally due to astronomical, meteorological and oceanographic forcing (as previously described for currents). However, there are also other elements at play that influence sea levels over longer time scales: for example, the impacts of geologically recent ice ages, where alternate periods of glaciation and warming have led to changes in sea level of more than 100 metres; and climate change, which is resulting in more recent sea level rises.

For water level there are two main components:

- tides
- surge

There are also two combined water level parameters:

- still water level – a combination of tide and surge
- extreme water level – a combination of tide, surge and wave crest elevation

(a) Tides

Tidal levels

Tides exhibit a range of different levels (or heights) related to the various daily, monthly, annual and decadal astronomical cycles of the Moon and the Sun relative to the Earth. This means that they are predictable for any given time, date and location.

The principle tidal levels are as follows:

- HAT – highest astronomical tide
- MHWS – mean high water springs
- MHWN – mean high water neaps
- MSL – mean sea level
- MLWN – mean low water neaps
- MLWS – mean low water springs
- LAT – lowest astronomical tide

LAT is the lowest level that can be predicted to occur under average meteorological conditions. Conversely, HAT is the highest level.

To give an indication of what to expect for each level, typical ratios related to HAT are provided here (note that these ratios are approximate and for guidance only):

- MHWS is 0.9 × HAT
- MHWN is 0.7 × HAT

- MSL is 0.5 × HAT
- MLWN is 0.35 × HAT
- MLWS is 0.15 × HAT

Tidal range

Tidal range is the vertical difference between two nominated tidal levels – for example, HAT and LAT. Note that in some locations the tidal range can be zero. These locations are known as amphidromic points.

From LAT to HAT is the extreme tidal range and occurs once every 18.6 years. Note also that HAT is not the highest sea level possible, nor is LAT the lowest. This is because surges can increase or decrease predicted tide levels (see (b) Surge, below).

Tidal cycles

Although the tidal regime can vary in different parts of the world, the key features tend to be dominated by daily and monthly cycles:

- A semi-diurnal cycle, resulting in two high and two low waters per day (actually a lunar day, which is 24.8 hours). Note that in some parts of the world the cycle is diurnal, with one high and one low water per day, whereas in other parts of the world a mixed version of semi-diurnal and diurnal occurs.
- A spring/neap cycle of 29.5 days (lunar month), with higher high waters and lower low waters at springs, and vice versa at neaps.

(b) Surge

Surge, sometimes referred to as storm surge, is a change in sea level (either positive or negative) due to meteorological (rather than tidal) forcing: for example, storms moving across a sea area. It is usually not predictable, although levels can be inferred just before a surge event. A surge can be positive or negative, i.e. it can increase or decrease the predicted tidal level.

(c) Combined water level parameters

Still water level (SWL)

Still water level is an engineering abstraction calculated by combining tide and storm surge but excluding variations due to waves. It is normally expressed relative to a datum such as LAT or msl.

Extreme water level (EWL)

Extreme water level is an engineering abstraction calculated by combining wave crest elevation, tide and storm surge (Figure 2.4). It helps to determine the cellar deck level on offshore installations This is the lowest deck level above the sea which needs to avoid wave inundation – for instance, Emergency Shutdown Valves (ESD) valves. The predicted EWL should always be below the cellar deck level. The vertical gap between the EWL and the height of the cellar deck is known as the 'air gap'. EWL is normally expressed relative to LAT or msl.

How these combined water level parameters are derived is discussed in Chapter 7.

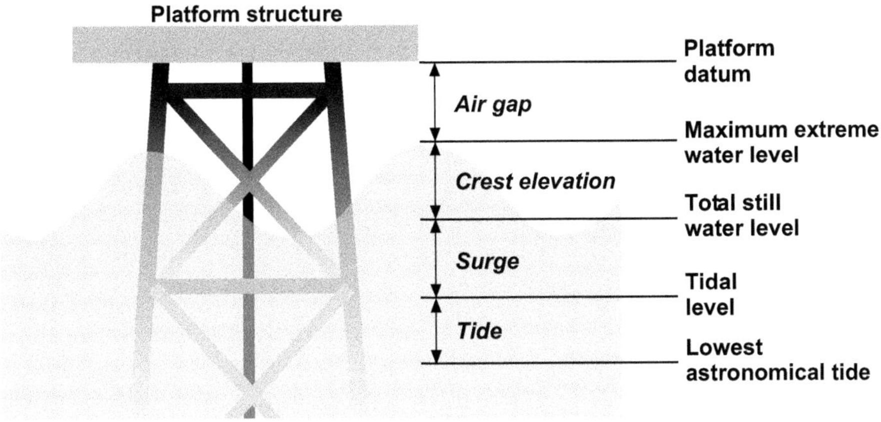

Figure 2.4 Schematic of water level terminology.

Units and conventions

Units

Metocean parameters are generally expressed in SI units as follows:

- heights, depths and levels in metres (m)
- periods in seconds (s)
- speeds in metres per second (m/s); speeds in knots are also permissible

However, Imperial units such as feet still tend to be the norm in the USA.

Directional conventions

Standard meteorological and oceanographic directional conventions are:

- winds and waves are 'from'
- currents are 'towards'

Thus a SW wave is from the SW, whereas a NE current is also from the SW (but flowing towards the NE). This means care is needed when analysing data and producing statistical outputs, and when end-users interpret the results.

Some users of metocean information prefer all parameters 'aligned', i.e. winds, waves and currents presented as 'from', so that interpreting and using the output is easier.

Wind elevation conventions

Winds are normally measured offshore at elevations up to, and sometimes in excess of, 100 m ASL. The globally recognised elevation standard or convention for wind speeds is to reference them to 10 m ASL. This means that any recorded wind data need to be reduced to this level unless there is a need to preserve the data at their measurement level: for example, data collected at different vertical levels for OR requirements.

As a rule of thumb, typical differences in wind speeds between 100 m and 10 m are 20%, i.e. speeds are lower at 10 m.

Current profile conventions

Current profiles are usually expressed as depths below the surface but they can also be referenced to the height above the seabed. Profiles may also be expressed as a ratio to depth, with 1.0d equating to the surface, 0.5d as mid-depth, and so on.

Chapter 3
The lifecycle of an offshore development and metocean deliverables

The project lifecycles for OR and O&G

The lifecycles of OR and O&G projects tend to be similar. They may be summarised in terms of six key stages.

For OR the six stages are:

(1) site selection and feasibility
(2) development and consents
(3) design and certification
(4) installation
(5) operations and maintenance (O&M)
(6) decommissioning

For O&G the six stages are:

(1) exploration (for hydrocarbons) including seismic, hydrographic and geotechnical surveys and drilling
(2) identification and assessment of potential developments, selection of potential concepts/structure type and final approval
(3) design and construction of approved concept/structure type
(4) installation
(5) production (of hydrocarbons) or operations and maintenance
(6) decommissioning and field abandonment

Note that, in stage 2 for O&G, 'concepts/structure type' is a generic term encompassing a potential wide range of solutions: for example, a fixed steel platform, a tension-legged platform (TLP), a floating production, storage and off-take (FPSO), subsea installation plus associated pipelines etc.

Management of either lifecycle not only requires achievement of targets for each stage but also a transfer of knowledge, practices and findings between stages, in particular between the design and operations teams because these are typically distinct and separate departments within an organisation.

For each stage there is a specific set of metocean requirements, which in turn needs a range of products and services, or deliverables. In general, as each stage is progressed an increasingly sophisticated metocean input is needed.

In order to satisfy these requirements, it is important to consider:

- early metocean input into the lifecycle
- establishment of a metocean focal point to co-ordinate activities
- an appropriate metocean contracting strategy
- continuity of metocean input throughout the lifecycle

Metocean deliverables

Chapter 1 introduced the importance of a co-ordinated and integrated approach to metocean activities, and the previous section of this chapter has outlined the stages in the OR and O&G project lifecycles. This section takes things a step further by outlining a recommended strategy that should be adopted to ensure that metocean decisions, requirements and products throughout the life history of a project are delivered optimally.

It is structured in terms of the key deliverables associated with the various stages of the project lifecycle for both OR and O&G. There are a considerable number of deliverables; however, it will be seen that some are updates as project work progresses over time.

(a) Deliverables for lifecycle stages 1 and 2

Deliverable 1: initial metocean briefing note

The first metocean deliverable, at the earliest stages of a project, should be a briefing note, indicating any known metocean conditions or issues that people need to be made aware of at this early stage. This can be expanded to include some basic statistics (if available), existing data sources and availability, plus any recommendations such as future data collection requirements and an outline metocean strategy.

Deliverable 2: preliminary metocean report (existing information)

This report covers the development of a range of statistical products based on existing available data. This would include a range of operational statistics (see Chapter 7) for planning and budgeting geotechnical, hydrographic and seismic surveys and for assessing potential downtime. For example, surveys generally require an Hs <2.5 m.

Production of initial and conservative 50-year or 100-year design criteria should also be included. However, it should be noted that in the early stages of a project there may not be enough data to produce meaningful statistics. This means care is needed in how such statistics are produced and presented.

Deliverable 3: weather forecasts

Weather forecasts are an essential deliverable at many stages of the project lifecycle. Their first requirement in the project lifecycle is to assist with the safety and execution of the various offshore survey operations: for example, identifying potential weather windows up to one week ahead.

Deliverable 4: data collection programmes

The collection of metocean data at an early stage in the project lifecycle is essential in order to gain a better understanding of conditions and to start building a database to assist with validating hindcast models (see below, Deliverable 5). This is to ensure that operational statistics and design criteria are more representative for any future development.

Collection of metocean data will include the deployment of various sensors, either remotely or fixed to a structure. This can include wind, wave, current and tide sensors.

The data can also provide real-time information to support survey or drilling operations.

Deliverable 5: hindcast data

Because there is a need to acquire long-term data for the derivation of accurate operational statistics and design criteria, it is important at this time to consider the requirements for a long-term (40+ years) hindcast database to contextualise potentially shorter measured datasets from deliverable 4, and thereby extend the data collection period.

It should be noted that a hindcast database may already be available as in the past such databases were developed in different parts of the world via Joint Industry Projects (JIPs) with multiple operators sharing costs. This potential availability needs to be checked.

Deliverable 6: updated metocean reports

As both measured and hindcast data become available, operational statistics and design criteria should be reviewed and, assuming enough data are available, should be updated. Such criteria should err on the side of conservatism until a long-term database is available. Long term in this context means decades rather than years.

Deliverable 7: input to PON15 (consent to drill) document (O&G only)

The PON15 drilling consent form requires a metocean summary paying particular attention to currents. The information provided for this may also be required for relief well planning, as part of the consent process.

(b) Deliverables for lifecycle stages 3 & 4

Deliverable 8: continued data collection and hindcast database

Data collection should continue to build a longer dataset for improved statistical products and for the verification of hindcast data.

Deliverable 9: final metocean design criteria report for Front End Engineering Design (FEED)

This is one of the most important deliverables as it will contain the best estimates of extreme metocean conditions for use in the final design. It usually requires a review and update, as appropriate, of any previous criteria provided in earlier preliminary/feasibility studies.

It should also be possible to reduce some of the conservatism in the earlier studies.

Deliverable 10: specifications for an installation-based metocean station (O&G only)

If the concept is a fixed or floating structure, specifications for an on-board metocean station should be prepared. This should include a listing of competent contractors, recommended equipment, sensor locations, recording requirements etc.

Historically, specifications have tended to be passed to the telecoms department or contractor. However, this is not the preferred route as such personnel may not be familiar with metocean standards and requirements, which can result in non-compliance with regulations. It is therefore better for O&G companies to deal with competent metocean contractors directly.

Deliverable 11: onshore metocean report (optional)

As structures are built onshore, there may be a requirement to provide a metocean report summarising onshore conditions at the location. This should focus primarily on meteorological conditions but there may be a requirement to include nearshore sea conditions as well.

Deliverable 12: tow-out metocean report (design criteria and operational statistics)

Towing a structure from its coastal construction yard to its offshore location requires a metocean report detailing design criteria and operational statistics for the tow-out route and any potential stop-off points or sheltered areas along the way in case of severe weather.

Some tows can potentially cover considerable distances and, as a consequence, involve travelling through areas of differing metocean conditions. For example, a structure may be installed in the Gulf of Mexico but be fabricated in the Far East and be towed from there to the USA. This will require substantial investment in metocean analysis and reporting.

Deliverable 13: installation site metocean report (design criteria and operational statistics)

The final metocean design criteria report, as per deliverable 9, should contain appropriate design criteria for the installation site. However, there will also be a requirement for detailed operational statistics to assist with installation planning, heavy lifts etc.

Deliverable 6 may be appropriate for this but it may require an update and, potentially, further analyses if additional data are available.

Deliverable 14: weather forecasts for tow route and installation site

Weather forecasts are essential for planning and executing tow operations and for identifying optimal routes and potential weather windows up to one week ahead. Tows only proceed when a favourable weather forecast is predicted: normally at least 3 days good weather with wave heights within tow limits.

Deliverable 15: real-time data for installation site

Real-time metocean data (especially waves) are essential, particularly with respect to jacket installations and heavy lifts. Continued data collection as outlined in deliverable 8 should suffice.

Deliverable 16: commissioning of on-board metocean station (O&G only)

The metocean station referred to in deliverable 10 should be checked and commissioned soon after platform installation, to ensure it is providing accurate information for daily operations.

(c) Deliverables for lifecycle stage 5

Deliverable 17: real-time data for daily operations

Real-time metocean data are required to assist daily operations such as crane activity, supply boats, helicopter operations and maintenance activities, and to further extend the metocean database. This deliverable will be required throughout the lifetime of the installation.

Deliverable 18: weather forecasts

Weather forecasts are essential for planning and executing operations and for identifying potential weather windows up to one week ahead. This deliverable will be required throughout the lifetime of the installation.

Deliverable 19: updates to design criteria and operational statistics

As more data become available it is important to review and update design criteria and operational statistics on a regular basis. As a minimum it is recommended this is done every five years and that updated reports are issued.

(d) Deliverables for lifecycle stage 6

Deliverable 20: metocean report for tow route to shore (design criteria and operational statistics)

This report will contain design criteria and operational statistics for the decommissioning location and tow route to the shore. There may also be a requirement to produce additional information for locations to shelter from during severe weather.

Deliverable 19 should provide useful information for the installation site.

Deliverable 21: weather forecasts for decommissioning location and tow route

This is the same type of deliverable as deliverable 14 above.

Deliverable 22: real-time data for decommissioning location

This is the same type of deliverable as deliverable 15 above.

(e) Some generic deliverables

The above listing of deliverables is a sequential format for a typical offshore development. However, it should be noted that there are also other deliverables that may be required from time to time irrespective of the sequence and/or at several times during the sequence. Examples of these are now provided.

Tidal height/current predictions

These can be useful, particularly for vessel operations and subsea activities in shallow-water areas and/or areas with strong currents. Identification of periods of slack water can be beneficial for optimising maintenance work.

Predictions can also be important in sea areas where tidal currents are strong, to optimise seismic survey shooting times and to minimise the 'feathering angles' between adjacent survey lines. Feathering Angle is a technical seismic term. It addresses the angle between a ships heading and the cable it tows. Strong currents can push the cable sideways making the angle larger.

Verification of weather forecasts

It is important to check continuously the quality and accuracy of weather forecasts by comparing them to measured data. This should be done on annually to assess trends or bias.

Data collection: maintenance report

There will be a need to maintain regularly the metocean sensors and equipment used to collect data via dedicated offshore visits. A maintenance report is required for each visit, detailing the work carried out.

Data collection: quality-controlled dataset

Any data collected should, as a minimum, be quality controlled by the contractor and stored in their archive or sent to the client's archive. However, it is better to have these data subject to independent quality control by a competent third party. This removes any potential conflict of interest.

Data collection: quarterly/annual data summary reports

After the quality control of data, it is recommended that reports are produced summarising statistically the data collected for the period. For longer-term data collection such reports should be quarterly and annual.

(f) Summary of metocean deliverables

Although there are in excess of 20 deliverables, they all fall into one of the six activities outlined in Chapter 1 and repeated here:

- Activity 1: acquiring data
    - that is already available
    - through measurement by metocean sensors deployed offshore
    - by developing numerical models to simulate past conditions
- Activity 2: providing real-time display and transmission of data from offshore measurements
- Activity 3: managing the provision of weather forecast services

- Activity 4: performing quality control checks on the data, archiving the data into a database and ensuring their long-term management and availability
- Activity 5: carrying out studies and statistical analyses of the archived data to provide two main types of output
  - operational statistics focusing on typical conditions
  - design criteria focusing on extreme conditions
- Activity 6: providing briefing notes and expert advice

More detail on these six activities as they relate to the deliverables is provided in Chapters 4–8.

Chapter 4

Metocean data acquisition

As mentioned previously there is a wide range of end-users requiring different types of metocean data and information. For example, people working offshore generally require real-time data, i.e. what is happening now, whereas those based onshore tend to require statistical information derived from long-term data to assist with planning operations or the design of an installation.

The availability of good quality historical and present data is therefore a key requisite for the various end-user requirements. Although this usually means that there is a requirement to embark on a data collection campaign, the first consideration should be to establish whether data are already available.

To this end, there are two options:

- review the availability of existing data (that have already been collected)
- embark on a new data acquisition campaign

For both options it is important to be aware that there are two main types of data: measured and hindcast. With regard to these:

- 'measured' refers primarily to data collected at an offshore location by deploying various sensors or measurement devices – for example, a wave buoy or an anemometer. Note, however, that it can also refer to satellite data
- 'hindcast' refers to historical data that that are generated by a numerical model

Existing data (and information)

Whilst some existing data may be available internally within an O&G company, there are also a number of external sources both in the commercial and public domains:

- technical papers and UK Admiralty publications
- the internet: for example, 'one-click statistics portals'
- institutes/national databases: for example, the UK Met Office, the British Oceanographic Data Centre (BODC) and the NCDC National Climate Data Centre (NCDC)
- historical and existing measured data collection programmes publicly funded by government and/or agencies and institutes
- historical and existing measured data collection programmes privately funded by commercial organisations, including O&G companies
- numerical model and satellite outputs

As, potentially, there may be several different sources, there might be considerable variation in data quality, duration, costs, availability etc. It is therefore important to consider these issues to ensure that the data are suitable. For example:

- a short-duration dataset will not be useful for deriving 100-year values
- a low data return or dataset with gaps will result in less accurate statistics
- statistics from the internet may not be accurate or appropriate

Nevertheless using already available data – with suitable caveats and conservatism built in – is sometimes all that is possible when providing an initial overview of metocean conditions. Even so, it should be noted:

- some end-users will require data in real-time
- there may be legal requirements associated with the need to collect data
- final design criteria for a development need to be based on a long-term site-specific dataset

Under such circumstances a new data acquisition programme will be required.

New data acquisition

If little or no existing metocean data are available then there will be a need to collect and/or to generate data by undertaking a measurement campaign and developing a hindcast database. In this respect it should be noted that an offshore development usually requires, and directly benefits from, a dedicated data collection programme.

A measurement campaign involves the installation of a range of metocean sensors, either by means of a remote mooring deployed in the sea or attached in some way to a vessel or installation. This is generally referred to as site-specific measurement.

Site-specific measurements provide reliable quantification of conditions at a location during the instrumentation deployment period but they are usually of much too short a duration to represent a reliable quantification of long-term trends and statistics.

'Long-term' can mean different things to different people. With respect to metocean statistical products 'long-term' means:

- for operational statistics: at least 10 years of wind and wave data and 2–3 years of current data are needed
- for the derivation of design criteria: 30–40 years of wind and wave data and at least 10 years of current data are needed.

In practice this means that, in the time available, it is often impossible to acquire that quantity of data by measurement alone. The solution is to use a numerical model to simulate past metocean conditions and generate historical data over a much wider area. This is referred to as spatial hindcast data (in contrast to the site-specific nature of measurements).

Such data can be generated relatively quickly for periods in excess of 40 years, and can be calibrated with, say, a couple of years of simultaneous measured data.

Although 40 years of data may sound excessive there is clear evidence of long-term trends and cycles in many of the metocean parameters. This subject is discussed further in Chapter 9 with respect to a long term wave database In the North Sea.

Site-specific measurements

Site-specific measurements are required for three main reasons:

- to provide data in real-time to support operational decision-making
- to build up knowledge of metocean conditions at a location and establish a long-term database for future statistical analyses
- to validate and calibrate longer-term hindcast databases

There are many issues to consider and to address with respect to a data collection programme to ensure that good quality measurements are obtained. These are summarised in turn.

(a) How to collect data and which sensors to use

There are three main ways of gathering site-specific measurements:

- the deployment of remote moorings anchored to the seabed
- attaching sensors to vessels and mobile installations
- installing permanent platform-based metocean systems

Moorings are normally utilised when there is no infrastructure in the area. Once an installation is on site, sensors and systems are better placed on or close to them.

Platform-based metocean systems for O&G are relatively commonplace, and in most parts of the world it is mandatory to include one on any new installation. With respect to OR the use of dedicated met masts – which are effectively mini platforms – is also becoming more common.

The choice of sensors to be used can depend on a number of different factors. Further information is provided later in this chapter in the sections on measuring winds, waves, currents and water levels.

(b) Sensors and equipment: buying versus hiring

It is normal practice to hire sensors and equipment from a metocean contractor. Whilst this is appropriate for short-term measurement campaigns, for long-term campaigns (i.e. those in excess of 1–2 years) it may be more cost effective to consider purchasing rather than hiring, using the contractor to maintain the equipment.

(c) When to start data collection

It is vital that data collection starts early enough to enable the collection of sufficient data. Consideration should also be given to the time of year. Deploying moorings in winter months, for example, can be more hazardous and take considerable time, thereby adding to costs.

(d) What parameters to measure

It is important to consider the specific parameters required in any measurement program: for example, Hs, Tz etc.

(e) Number of locations and data sampling

The number of locations where measurements should be made will depend on the size of the site or area in question, the complexity and variability of the oceanographic conditions in the region, and the existence and availability of prior metocean measurement data.

Knowledge of expected conditions and phenomena in an area or region is also vital for ensuring that metocean data are sampled optimally both in terms of frequency and duration. Some of the processes or phenomena that can have implications for the number of locations and sampling regime may be:

- high-frequency currents (changing speeds rapidly in minutes)
- squalls and rapidly changing wind speeds
- the presence of eddies

There are three key terms associated with how data are sampled: sample duration, sampling frequency, and recording interval.

Sample duration

This is the period of time over which data are measured, and can range from seconds to hours, depending on the parameter. For example, winds are often sampled over 10 minutes.

Sampling frequency

This is the amount of times that a parameter is sampled within the sample duration. For example, with respect to wind in a sample duration of 10 minutes, the sampling frequency could be every second, generating 600 individual values. Sampling frequency may also be expressed in terms of Hertz (Hz), where 1 Hz is defined as 1 cycle per second. Thus a 2 Hz sample equates to every 0.5 seconds

Recording interval

This is the time period between one sample duration and the next. For example, if the sample duration is 10 minutes every hour, then the recording interval is hourly.

It is very important to ensure that the sampled data are accurately capturing all the different metocean processes applicable to the area of interest. For example, sampling hourly or averaging over an hour would miss high-frequency currents or wind gusts, resulting in deficient datasets.

It is likely that potential contractors will be able to provide recommendations as to the optimum campaign design. Accordingly, OR and O&G companies are encouraged to discuss in depth with potential contractors their requirements and eventual intentions for the use of the data.

(f) Quantity of data needed

This subject has been touched on earlier in the chapter in the section on New Data Acquisition. However, one additional point to add is that consideration needs to be given to the continuation of data collection into the production phase of a development, in order to:

- provide information for future operations and maintenance (O&M) activities
- extend the metocean database and thereby improve the accuracy of statistical analyses

(g) Real-time data access versus recorded data

It is important to consider real-time access to data (as well as recording the same), not only for operational decision-making but also for monitoring and quality control purposes. See also 'Data and equipment loss' below.

(h) Data and equipment loss

The deployment of sensors and equipment by means of remote moorings in a hostile marine environment can sometimes lead to a loss of sensors and data due to storms, vessel collisions, fishing activity and vandalism. Consideration must therefore be given to the level of data loss deemed acceptable.

In this respect it is important to ensure that any contract incorporates risk management and emergency response in order to rectify problems, as well as a strategy to minimise data loss. Data loss can be mitigated by:

- real-time transmission of data
- redundancy in the measurement campaign by installing multiple moorings
- regular servicing of equipment (say 3 monthly or 6 monthly)
- informing relevant maritime organisations of the measurement campaign

As data loss is always a potential issue, consideration needs to be given to the contractual implications and liabilities if data losses reduce overall data returns to unacceptable levels.

(i) HSE issues

It is important to appreciate that activities associated with site-specific measurements are inherently hazardous. In this respect it is vital that sufficient attention is paid to health, safety and environmental issues. Some examples include:

- deploying and recovering equipment from the stern or sides of vessels
- working at height on offshore installations
- the potential for dropped objects falling from height

Personnel to be used in offshore operations must possess the correct certification (for example, offshore survival training), and vessels must be checked for their suitability in deploying and recovering equipment and to ensure their compliance with health and safety requirements.

More detailed information can be found in document OGP 447 (see Appendix 1).

(j) Data processing, quality control and reporting

There are five key points to consider with respect to data processing, quality control and reporting:

- processing of data should follow recognised industry approaches
- it is recommended that raw, as well as processed data are retained for future use – for example, when investigating storms in detail
- ensure data are quality controlled on a regular basis, ideally monthly or quarterly
- a metocean measurement campaign can generate large quantities of data. It is necessary to plan in advance how and where data will be stored and archived, and how such data will be accessed.
- ensure quarterly/annual statistical analysis reports are produced summarising the data in appropriate tables and graphs

Data quality control issues are discussed further in Chapter 6.

(k) Contract process

Once budget approval has been given to proceed with a data collection campaign there are a four key points to consider with respect to the contract process:

- produce a fit-for-purpose scope of work (SoW) and detailed specifications. These will contain inter alia equipment, personnel, analysis and reporting requirements.
- assess contractors competent to carry out the work
- incorporate the SoW and specifications into the data collection tender documents and issue to contractors
- evaluate tenders financially and technically, and award contract

Note that lead times for the contracting metocean equipment and for its deployment may vary from contractor to contractor. With this in mind, tenders should be issued in a timely manner (ideally at least 3 months before the contract starts), to allow potential contractors sufficient time for the acquisition of necessary licences and to ensure the availability of equipment and personnel.

Measuring winds, waves, currents and water levels

Although the following section goes into more detail on the specifics of measuring winds, waves, currents, water levels and other data types, it does not necessarily provide detailed specification information on all sensors.

Measuring winds

Measuring wind speeds and directions offshore can be quite challenging whether utilising a remote mooring or an existing installation. For example:

- using a remote mooring or a buoy can affect the accuracy of the measurements due to the motion of the buoy itself and the proximity to the sea surface
- using an installation such as an offshore platform or met mast can also affect the accuracy depending on the sensor location, as the structure itself can distort the wind flow

The preferred solution for O&G is to use an existing installation but to site the sensor in a location as free from flow disturbance as possible: for example, at the top of the derrick. However, consideration needs to be given to ease of access for maintenance.

For OR there is sometimes a requirement to install a dedicated met mast and position a number of wind sensors at different elevations

There are two main types of wind sensor: traditional or standard anemometers, which are typically mechanical with a rotor and vane, and acoustic or ultrasonic types, which have no moving parts.

The rotor and vane anemometer is still widely used but it is increasingly being replaced by the acoustic or ultrasonic sensor (Figure 4.1). This is because the latter is much lighter and therefore easier to install and maintain. However, it can be susceptible to icing without an internal heater, which can affect data quality.

Figure 4.1 Anemometers: left – rotor and vane; right – ultrasonic (courtesy Gill Sensors & Controls Limited).

Data are normally sampled every second, and from this a range of mean values and gusts is derived, typically 1-hour mean, 10-minute mean, 1-minute mean and 3-second gust.

Most recently a different new-technology approach to wind measurement by means of LiDAR (light detection and ranging) has become more widely available, although its use remains rather limited at present.

LiDAR began as a ground-based remote sensing tool, which although normally sited onshore could also be installed on offshore installations. It employs the Doppler effect (electromagnetic radiation is reflected off particles) to detect the movement of air in the atmospheric boundary layer (ABL) from which the wind speed and direction can be inferred. More recently, floating LiDAR systems have also become available as moored buoys giving wind information up to 200 m vertically above the sea.

Measuring waves

The measurement of waves can also be quite challenging. As with winds there are two main methods or sensor types: a remote wave buoy moored to the seabed, and platform-based sensors that generally 'look down' and scan the sea surface (Figure 4.2).

Buoys used to be the standard for wave measurement but, over the last twenty years or so, platform-based sensors have become more common.

Wave sensors used to be capable of measuring only the height and period, with no directional information. Such sensors are referred to as omni-directional. In the 1980s and 1990s directional capabilities were developed for both buoy and platform-based wave sensors need to be sited as far away as possible from the potential effects of data contamination by waves interacting with an installation's legs or monopole. Today both omni-directional and directional sensors are in use – the selection of whether

Figure 4.2 Measuring waves: wave buoy (left, courtesy EMEC); platform wave sensor (right, courtesy RS Aqua).

to go for a directional sensor (which is more expensive) will depend on information requirements and budget availability.

There is also a decision to be made with respect to sensor type, i.e. buoy or platform-based, as both have advantages and disadvantages.

The main issues with buoys are:

- safety concerns associated with deployment and recovery of the buoy during installation and service visits
- potential 'skirting' of crests due to mooring constraints
- fishing and other vessel activity causing damage or loss
- damage or loss during storms
- potentially lower data returns than platform-based sensors

The main issues with platform-based sensors are:

- finding a suitable location, as platform-based sensors need to be sited well away from the potential effects of data contamination by the platform legs
- the need for considerable cabling from the sensor location to the metocean processing unit, which may be located in the control room
- the potential for accidental damage to the sensor/cabling by platform staff

History of wave measurement and processing

As waves are usually the most important metocean datatype, it is important to understand the history of wave measurement because there have been many changes in sensor type, processing methods and parameter derivation over the last 50+ years due to technological advances. These can have implications for the derived parameters and subsequent analysed products such as design criteria (discussed in Chapter 7).

The following sections summarise the history of the various changes and discuss their implications, paying particular attention to Hs and Tz.

1960s to mid-1970s

The first wave-measurement programmes for the O&G industry began in the late 1960s in the North Sea. In those pre-computer times wave data were recorded on to paper chart rolls (referred to as 'chart records'), as a continuous ink trace, which in essence constituted a time series of data.

It was assumed that wave conditions effectively remained stationary over a 3-hour period; hence waves were sampled for 20 minutes every 3 hours.

Each 20 minute chart roll was analysed in two phases: first carrying out visual counting and inspection of the wave records to obtain

- the two highest crests
- the two highest troughs
- HmaxT (the highest individual wave crest to trough)

and second the derivation of two key parameters:

- Hs (sometimes referred to as H1/3)
- Tz (the average zero up-crossing wave period)

Three different types of wave sensors were used at this time:

- a wave staff (attached to a structure such as a platform leg)
- a ship-borne wave recorder (SBWR)
- a datawell waverider buoy

Chart record data from this period are still available: for example, the early North Sea data collection for the Forties and West Sole platforms.

Mid-1970s

During the 1970s, as a new generation of computers appeared, it became a realistic proposition to gather data using computers. Instead of using paper chart rolls and pen-and-ink plotters, the signal from the wave sensor could be directed to a computer that recorded data to analogue magnetic tape. These tapes could then be shipped ashore and re-played into an analogue-to-digital converter. The technique, whilst an obvious first step, was fraught with problems. For example:

- tape-stretching spoiled the accuracy of the timekeeping
- tapes could only be re-played about 10 times faster than real time and re-play often required several attempts before a satisfactory result was obtained
- mechanical problems caused some recorders to rotate at uneven speeds, causing a cyclical wavelike corruption of the signal

Late 1970s to early 1980s

Processing power increased to the point where the signal could be processed offshore and only the derived parameters shipped back to the beach. However, the task was demanding for the central processing units (CPUs) of the day. Therefore the measurement regime was configured to gather data for approximately 20 minutes, and then process and record it to tape over the subsequent 40 minutes. In order to accomplish the task a new Hs parameter –H4rms – was introduced, being 4 times the root mean square (rms) of the wave signal. It replaced HsTD as the most common measure of wave height.

At this time cassette tapes were also introduced and, whilst these could hold large quantities of processed data, they were routinely shipped at monthly intervals so that system performance could be monitored.

Waverider buoys were the main sensors used, as SBWR's and wave staffs were phased out. Other buoys also began to appear at this time.

Mid-1980s to late 1990s

Advances in computer technology and processing techniques saw the introduction of digital systems and a completely new way of analysing wave data: spectral analysis. This technique enabled more detailed analysis of the data, looking in particular at the energy at different frequencies and also deriving more parameters: for example, the peak wave period Tp.

The technique is also referred to as frequency-domain analysis, to differentiate it from the chart-records approach, which is referred to as time-domain analysis.

Increased computer power and storage capacity also meant that sampling could be done more frequently at hourly intervals. Sampling itself was normally at a frequency of 2Hz, i.e. every 0.5 seconds. Frequency–domain analysis requires the number of data points in the sample to be a power of 2. In order to try and ensure the sample duration was similar to time-domain analysis, i.e 20 minutes, the number of data points or samples is 2,048 which equates to duration of 17 minutes and 4 seconds.

Although buoys were still widely used, new types of platform-based sensor were introduced and over time began to replace buoys at many locations. Also, during this time, directional wave sensors (both buoy and platform-based) were developed and deployed at a few locations.

Late 1990s onwards

Over the last 20–30 years further enhancements to computing power and capacity have enabled continuous recording and storage of raw data, which enables retrieval of information for detailed investigation of individual events. Waves are still sampled over 17 minutes and 4 seconds (but sometimes for double this time based on 4,096 samples) and updated every minute on a rolling basis and stored accordingly.

Sensors are now dominated by platform-based ones although buoys still play a role. The dominance of platform-based sensors is primarily driven by safety considerations and the potential for enhanced data returns.

Implications of changes to wave measurement and processing

There have been several studies investigating the impact of changes in sensor type, processing methods and parameter derivation. One of the most interesting studies was carried out by Shell in the 1980s. It looked at the impact on the mean wave period (Tz) arising from the three different recording and processing techniques referred to above: namely, chart records, analogue tape and digital systems.

Wave data in the Brent area of the northern North Sea were recorded and processed using the three techniques and, after analysis, the following results were obtained for the average annual Tz for each:

- chart records 7.5 secs
- analogue 5.5 secs
- digital (spectral) 6.5 secs

These variations may not seem large but, over a period of one year, they have major implications for the number of waves hitting a platform. For example:

- 7.5 secs = 4.2 million waves/year (chart records)
- 5.5 secs = 5.7 million waves/year (analogue)
- 6.5 secs = 4.9 million waves/year (digital (spectral))

The maximum difference between the three techniques is 36%. This is clearly a major concern for structural engineers with respect to fatigue issues. For a platform with, say, a 30-year life a 36% decrease or increase equates to c. 45 million waves! It is therefore important to know the provenance of any wave data used in analytical studies. Clearly, statistics derived from chart or analogue data will give different results to those derived from digital methods.

As a postscript to the Shell study, metocean experts recommended the digital approach should be the standard.

Other studies have looked at the impact on the values for Hs derived from different sensors and the differences between Hs formulated as HsTD, H4rms and Hm0. In general these have shown differences of +/- 5%, which in most cases is not statistically significant.

Measuring currents

Whereas winds and waves are routinely measured at a number of locations the same does not apply for currents. This is primarily because:

- the behaviour of currents in tidally-dominated shelf seas (such as the North Sea) is reasonably well understood and there are often several sources of modelled current data available for such areas
- cost, safety and data-return issues are of concern as the measurement of currents often requires the deployment of multiple sensors at a range of depths via remote moorings
- waves and winds are generally more important because they usually present the largest loading for an offshore installation

However, the potential complexity of currents in shallow-water locations and in deeper waters off the continental shelf means that current measurements are more commonplace there, particularly in the latter for input-to-riser design and mooring analyses, and riser monitoring.

The traditional approach to measuring currents is by installing current meters at a number of fixed elevations on a mooring anchored to the seabed. These sensors provide speed and direction information at set intervals, normally every second averaged over 1, 10 or 20 minutes.

Historically these current meters were of a 'cup-and-vane' type (Figure 4.3) for quantifying speed and direction. However, over the last twenty years acoustic current meters have tended to be the preferred option.

Although traditional current meters can be used to derive a profile, considerable numbers would be needed on a single mooring. This led to the development of an acoustic sensor that could derive a profile from just one sensor: the acoustic Doppler current profiler (ADCP) (Figure 4.4).

This sensor resembles a giant torch in that it has four transducer heads that emit acoustic beams either vertically upwards or downwards depending on how they are deployed.

The beams require 'back-scatterers' such as plankton, and using the Doppler shift the speed and direction can be derived to give a complete current profile every 5–10 m

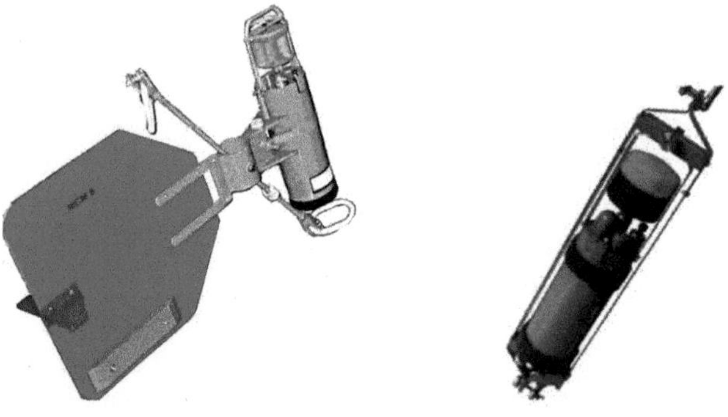

Figure 4.3 Cup-and-vane (left) and single-point acoustic (right) current meters.

Figure 4.4 Acoustic Doppler current profiler (ADCP) (courtesy Teledyne Marine).

throughout the water column. This means that, depending on the water depth, a single ADCP can provide the same data as installing, say, 20 or 30+ traditional current sensors.

ADCPs are available in different frequency versions, which relate to the depth penetration of the beams. For example, a 75 kHz version can provide current profiles in water depths of up to 600 m.

Although the sensor is most commonly used in upward- or downward-looking mode, placed either on the seabed or on a mooring, there are also versions that can be incorporated into the hulls of vessels.

Despite the fact that ADCPs are able to provide a wealth of information on current profiles, it should be noted that there are limitations to the availability and quality of data near the surface and near the seabed. Nevertheless they are widely used, particularly in deeper waters to assist with drilling operations.

As with traditional methods, sampling is usually once per second and processed as 1-, 10- or 20-minute means.

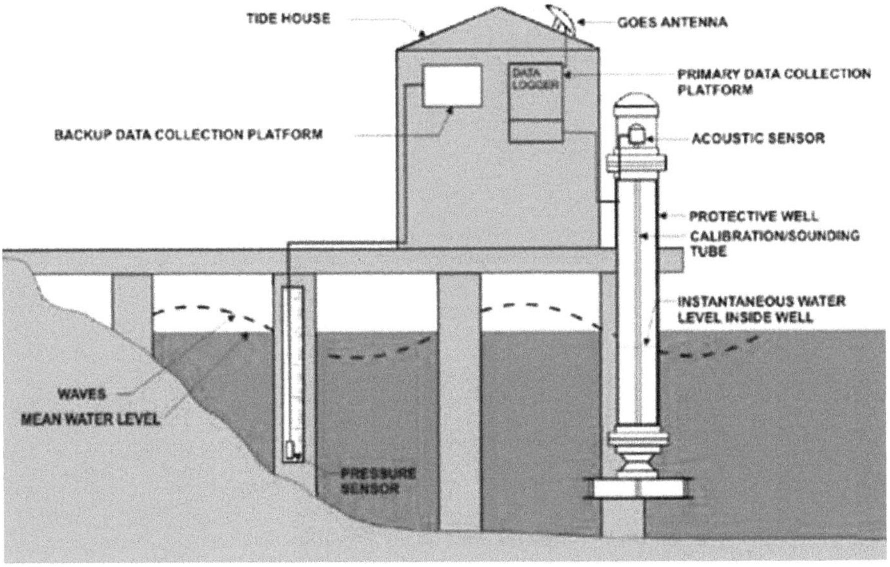

Figure 4.5 Schematic of a tide gauge system.

Measuring water levels

Water levels are not routinely measured offshore for reasons similar to those mentioned for currents, particularly as tides are predictable. Nevertheless many types of tide gauge sensor exist, which operate using different measuring principles. These include:

- tide poles or tide staffs
- mechanical float and stilling-well gauges
- pressure gauges
- bubbler gauges
- underwater pressure transducers
- acoustic gauges

It should be noted that platform-base radar wave sensor devices can, by virtue of their measurement technique, provide information on water level as well.

Figure 4.5 illustrates a tide gauge system.

Measuring other data types

Other meteorological

Other meteorological parameters – such as cloud height/cover, visibility, air temperature and pressure – are routinely measured at many offshore installations.

Such sensors are typically deployed close to helideck levels as most of this information is required in real-time for helicopter operations. Figure 4.6 shows an example of visibility and cloud sensors together with a Stevenson screen, which houses the pressure and temperature sensors.

Figure 4.6 Visibility and cloud sensors plus Stevenson screen (courtesy of Nevis Technology, Aberdeen).

Other oceanographic

Other oceanographic parameters, such as sea temperature and salinity, are not normally measured separately but rather can be incorporated into certain current and wave measurement sensors or mooring systems.

Satellite measurements

Satellites provide considerable quantities of data and information for a wide range of scientific and technical uses. A few of these satellites are providing metocean information from a range of dedicated sensors, which include:

- synthetic aperture radar (SAR) for measuring waves
- scatterometer (winds)
- radiometer (sea surface temperature)
- altimeter (sea surface height, currents)

These data can be classified as measured but, in contrast to site-specific measurements, which are effectively a single point, satellite data cover a wide area or 'swathe' as a result of the satellite's orbit around the Earth.

Although a considerable quantity of metocean data are generated spatially by satellites, it is not widely used by the OR and O&G communities, primarily because of 'repeatability' issues, i.e. the satellite positions and orbits mean that the same area of sea is passed over (sampled) only once every few days compared to the hourly or 3-hourly sampling of site-specific measured and spatial hindcast data.

This means that satellite data should be viewed as a supporting tool, particularly for spatial requirements, but should not be routinely used for metocean applications in OR and O&G. Satellite data can, however, provide indicative 'climatology' (metocean statistics) in areas where other data sources are limited.

Nevertheless there is still a wide range of end-users of satellite data in the broader weather and ocean community. In particular, satellite data are used as an aid to weather forecasting and for assimilation into numerical models.

Hindcast data

Site-specific measurements have been the standard approach for data collection for many years but the use of hindcast data is becoming increasingly common.

A hindcast is essentially synthetic historical data (such as winds and waves) generated by running a numerical model. In simple terms, it is possible to use past wind fields and barometric pressures to drive a wave model that can produce wind and wave data for each grid point on a pre-determined grid over large sea areas – for example, the North Sea or the Gulf of Mexico.

The generated (hindcast) output means that there can be more than 40 years metocean data for potentially hundreds or thousands of grid points depending on the model resolution.

Apart from the need to provide good input wind and pressure data, such models also need good bathymetry and reliable physics to ensure accuracy of output. There will also be a need for some site-specific measurements to ensure the model is well validated.

Thus, hindcast refers to historical data as opposed to a (weather) forecast that predicts the (short-term) future. However, it should be understood that hindcast data are used primarily for future predictions by analysing the data to produce operational statistics and design criteria.

There are two key reasons why the development of hindcast databases is becoming more common:

- collecting measured data is expensive, particularly if a number of different locations is needed
- it takes 10 years to get 10 years of measured data; it can take about 10 weeks to get 10 years of hindcast data

The numerical models themselves vary in their coverage, size and resolution, ranging from global to regional to local. Global models typically have a resolution of, say, 20 km, regional models typically less than 5 km, and local models typically less than 1 km.

Whereas global models are useful for giving broad overviews and picking out large-scale features, for OR and O&G applications the coarseness of their grid resolution is

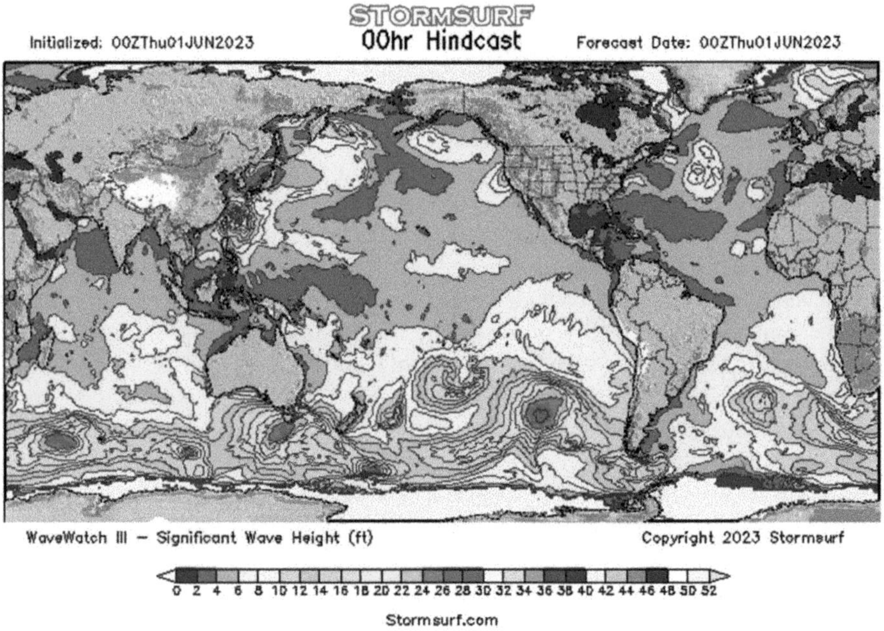

Figure 4.7 Global hindcast output of Hs (courtesy Stormsurf).

insufficient to successfully provide accurate information on a regional or local basis. In such cases regional models tend to be the norm – apart from nearshore requirements, where local fine-grid models are usually required.

Hindcast data output

Figure 4.7 is a snapshot of global hindcast output of significant wave height (Hs). It demonstrates the power of such models to show the variability of conditions on a global basis. But it also shows it is too coarse for providing specific values for regional sea areas such as the North Sea or the Gulf of Mexico.

It should be noted that although wave models are well advanced and provide good information, current models (particularly those for the open ocean) are less well developed.

Hindcast data are effectively a long time series of past information and usually consist of hourly or 3-hourly outputs of winds and waves for each grid point. In some cases currents and water level data will also be available. With respect to the former this may not always be a 3D model giving data at different depths, but a less sophisticated 2D current model effectively providing one value referred to as a depth-mean integrated current.

With a typical hindcast dataset consisting of 40+ years for potentially hundreds or even thousands of grid points, the vast quantities of data generated mean that there is a requirement for an extensive archive.

Most of the major shelf and enclosed seas of the world have already had regional hindcasts produced, usually by the O&G industry. These have tended to be organised

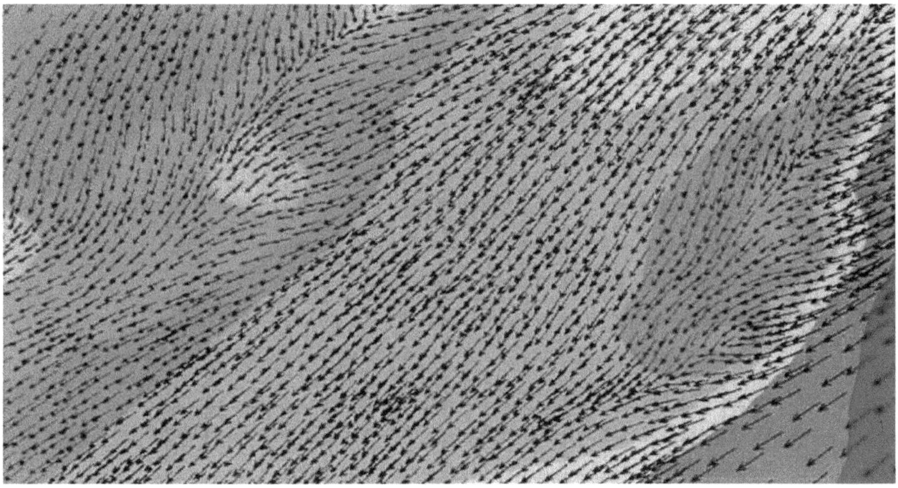

Figure 4.8 Output from a shallow-water model.

and funded via joint industry projects (JIPs). Examples include the NEXTRA project in the North Sea and SEAMOS in the South China Sea.

These existing hindcasts are usually confidential to the JIP participants. However, individual grid points can often be purchased by non-participants subject to the terms of particular JIPs.

Additionally, there are other hindcast databases available from a range of providers, including metocean contractors, institutes and governmental organisations.

Hindcasts and shallow water

As already mentioned, hindcast models that cover major ocean and shelf seas are not appropriate for shallow water (<50 m depth), coastal zones and areas of complex bathymetry. In such circumstances the use of a local hydrodynamic model is essential.

Such models enable much finer resolution – down to <100 m – in order to provide a more accurate representation of the metocean conditions in the area of interest.

Figure 4.8 shows an example of shallow water model output with wave heights in colour, and current speed and direction (arrows) on a 50 m grid. It illustrates the wealth of detailed information that can be generated with such a fine grid.

As with all hindcast models there is still a need for offshore reference data and local measurements within the model area to ensure proper model validation.

Measured and hindcast data: benefits and differences

Whilst there are considerable benefits associated with having both site-specific measured data and spatial hindcast data, it should also be appreciated that there are differences between them. This section quantifies the benefits for both types of data and highlights the implications of the differences.

Benefits of site-specific measured data

The benefits of site-specific measured data can be distinguished according to those applicable to real-time data and those applicable to recorded or historical data.

For real-time data the benefits of site-specific measured data are:

- site specific
- help with operational decision-making
- contribute to the safety of operations
- help to improve weather forecasts (if the data are sent ashore promptly)
- provide regulatory compliance

For recorded data the benefits of site-specific measured data are:

- contribute to 'incident analysis' investigations: for example, what the wave height on a particular day in the past was
- provide validation of numerical models
- enable verification of weather forecast accuracy
- enable identification of individual waves (from raw data), which can be important in high-sea states
- help to build up a long-term metocean database (providing that the measurement campaign continues), in order to
  - o derive and refine operational statistics and design criteria
  - o assess any trends in the data

Benefits of hindcast data

Some of the benefits of hindcast data are also similar to those for recorded data: for example, 'incident analysis' and building up long-term databases. However, there are additional benefits. Hindcast data:

- allow for considerable expansion of available data both spatially and temporally. This should facilitate
  - o more confidence in any derived statistics
  - o a better assessment of any trends and cycles in the data
  - o infilling of 'spatial gaps' between two or more measurement locations to allow an assessment of the spatial variability of metocean conditions between them
- will be more homogeneous compared to measured data, i.e. the same model will generate 40 years of data whereas, as shown previously, wave measurement techniques have changed over the last forty years
- models have the ability to capture 'processes' and information for areas of complex oceanography/bathymetry; but only when run with optimal grid resolution
- provide a quicker way to obtain a long-term dataset
- are more cost-effective than measurements: for example, a 40-year hindcast is cheaper than 1 year of site-specific measurements at one location. However, there is still a need for site-specific measurements to validate the hindcast data.

Differences between measured and hindcast data

Whilst recognising the prudence of having both measured and hindcast data, it is important to appreciate the potential differences between the two types of data. There are three key differences worth noting:

(1) site-specific measurements are carried out at a location or point, whereas hindcast data are a spatial average over say 5 × 5 km
(2) measurements are made by sensors (which may seem to equate to the 'real world'), whereas hindcast data are generated by a computer (which may be perceived as 'synthetic')
    BUT it is also worth noting that there can be differences in output data between two or more sensors (whether they are of the same type or different types), and two or more models for the same location.
(3) a time series of measured data appears 'rough' when inspected visually because of frequent sampling (every 20 minutes), whereas hindcast time series appear 'smooth', due to less frequent sampling (hourly or 3 hourly)

These differences can result in different output values for simultaneous data, which may have implications for derived statistics.

Presentations that compare measured and hindcast data can help to illustrate these differences. For example:

- time series plots may show that simultaneous measured and hindcast data can look similar; on closer inspection there may be slight offsets and differences in peak values during storms
- regression and/or quantile analyses can quantify any offsets and show whether measured or hindcast data are higher or lower
- joint frequency tables, such as wave scatter diagrams of height versus period, can show any differences or offsets in wave period
- extreme value analyses (EVA) can result in different best-fit slopes for each type of data and hence different extremes will be derived. EVA is explained further in Chapter 7

It is, therefore, important to understand and appreciate not only the benefits of both measured and hindcast data but also the differences between them, so that data accuracy and relevance is assured and derived statistics are meaningful and appropriate. At a minimum this means ensuring that comparisons are made between available measured and hindcast data, and that competent metocean expertise is utilised.

Chapter 5
Weather forecasts

Weather forecasts are a routine daily requirement for OR and O&G companies and for the contractors that support the various operational and design requirements.

The subject of weather forecasting is obviously a wide one in its own right and worthy of a separate book. It is therefore not my intention here to provide a detailed account of the subject but, rather, to focus on general issues relating to OR and O&G requirements, as well as some specific products provided by forecast providers.

It is important to appreciate that the type of forecast available on television or free on the internet is not appropriate for the specific and detailed requirements of the OR and O&G sectors. Indeed, such forecasts will not be accepted, for example, by regulators, warranty surveyors and certifying agencies. Consequently, weather forecast providers have developed a range of services and products specifically dedicated to the offshore community.

These services and products include:

- site-specific and operation-specific forecasts
- hourly forecasts with frequent updates (usually 6 hourly)
- standard forecasts looking ahead 3–5 days
- longer-term forecasts up to 2 weeks ahead, usually by means of ensemble or probability-based methods (see next section in this chapter)
- a wide range of forecast output in both tabular and graphical formats
- a wide range of delivery methods, including web-based services and smartphone delivery as well as traditional methods such as email
- 24-hour advisory services and office briefings
- verification of forecast accuracy

With respect to the weather forecast providers themselves, there are primarily two types of organisation:

- state, governmental or quasi-governmental organisations; usually one per country
- commercial or private companies, either exclusively focusing on forecasting or providing forecasting along with other metocean services

It is important to stress that the data and information obtained and used by weather forecast providers to produce forecasts varies. Furthermore there are differences in software, techniques and manual intervention techniques for each provider. The consequence of this is that the accuracy of forecasts varies between providers and can result in differences in forecast outputs and values for the same area/location. This should lead OR and O&G companies to give careful consideration when selecting a forecast provider. This may mean carrying out a pre-tender comparison study of forecast outputs for a number of different providers.

Some OR and O&G companies may occasionally obtain forecasts from two different providers as a way to get around this. This is particularly so for critical, weather-sensitive operations. However, using two different forecasts may prolong the duration of any activities if they rarely agree.

Probability-based (or ensemble) forecasts

Traditional forecast products provide a best estimate of future conditions based on a most likely scenario. In this respect they will state, for example, that the Hs will be 3.5 m at 12:00 on Day 2. Because it is a best estimate there is no additional information on the likelihood of Hs being say 2.5 m or 4 m. Such forecasts are also referred to as 'deterministic'.

This has led to the development of so-called probability-based forecasts, which indeed do expand on the best estimate by providing probability levels for the most likely value less or more than best estimate or most likely. This is achieved by running forecast models multiple times with each run having slight changes in input, which results in changes in output. As a result there will be many potential outcomes for the Hs at 12:00 on Day 2.

Output can be presented in a number of different ways. Figure 5.1 shows a range of values for Hs looking ahead 10 days in a time series format. Each line represents the output from a different model run, with the dark blue line presenting the median (or most likely) result from the runs.

There is reasonable agreement over the first few days, as would be expected. For example, the plot indicates a median value of 1.3 m at 15:00 on 21 July with a range 0.7–1.4 m. However, towards the end of the period more variability becomes apparent, and by Day 10 there are significant differences between the lowest forecast Hs (0.2 m) and the highest (4 m).

Figure 5.2 shows 10-day forecast information for wind speed presented in bar-chart form as probability levels. This plot is also referred to as a meteogram.

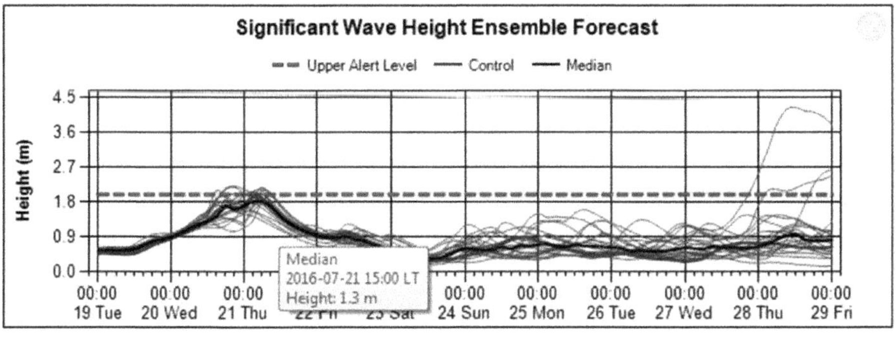

Figure 5.1 An example of a time series ensemble forecast output for Hs.

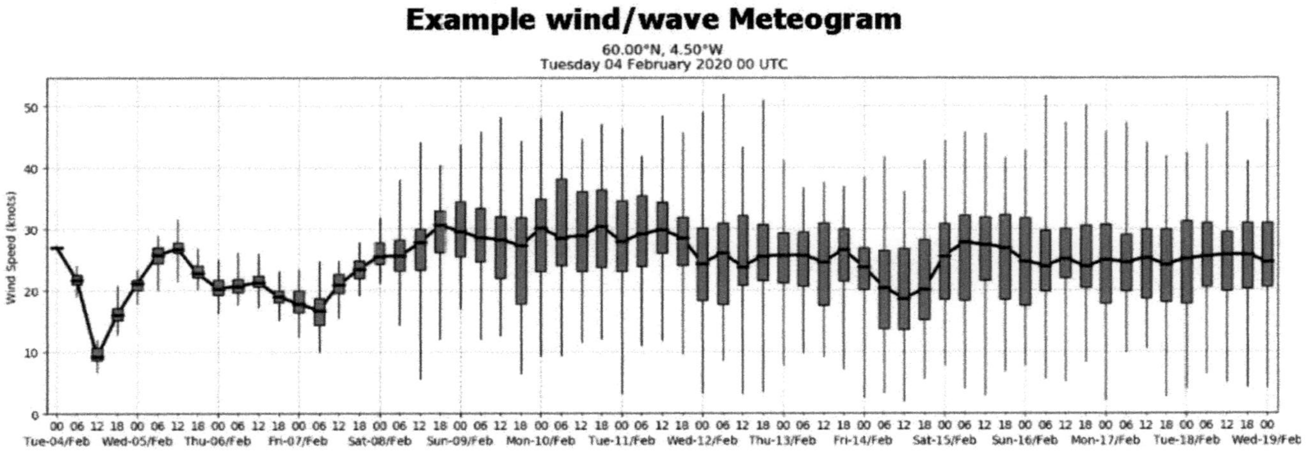

Figure 5.2 A meteogram ensemble forecast output for wind speed (courtesy of the Met Office. Contains public sector information licensed under the Open Government Licence v3.0).

There are three key pieces of information presented:

- the black line is the median (most likely) predicted value
- the rectangular 'box' represents the 25–75% probability
- the vertical lines provide the 0–100% probability

Thus on Saturday 8th Feb at 12:00:

- the median wind speed is circa 27 knots
- the 25–75% probability is circa 23–30 knots
- the 0–100% probability is 5–45 knots

The additional information from ensemble forecasts can often help with operational decision-making, particularly for weather-sensitive operations 7–10 days ahead.

Ocean forecasts

It is worthwhile mentioning here that, whilst most people are familiar with weather forecasts, ocean forecasts (primarily currents) are less well known.

Forecasts – or more correctly predictions – of currents in tidally-dominated areas or seas are readily available. However, for areas with weak tidal currents and/or dominated by oceanic influences, providing forecasts is more challenging due to the complexity and unpredictability of currents which existing models cannot always replicate well.

This means that it is not presently possible to get, for example, accurate hourly forecasts of current profiles for such areas. However, matters are improving as oceanic models and knowledge improve.

Figure 5.3 provides an example of an ocean-model output showing detailed sea temperatures for the North Atlantic visualised via Google Earth.

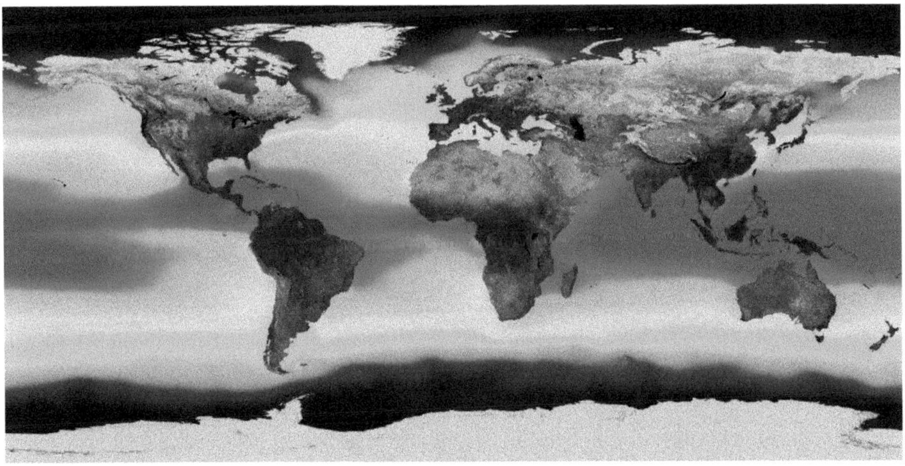

Figure 5.3 A map of global sea surface temperatures (courtesy of NASA).

Chapter 6
Metocean data quality control and archiving

Chapter 4 described the acquisition of measured and hindcast metocean data. The accuracy and quality of such data are vital in ensuring that end-users can make confident decisions on operational and design issues. This can be achieved by:

- quality control of the data
- data management and archiving

Addressing these issues enables the value of the initial metocean investment to continue into the future.

Hindcast data have normally been through a series of quality control checks and verifications by the contractor and, when supplied, are normally ready to use, although some end-users may wish to perform a few of their own checks by reviewing analytical plots and/or comparisons with measured data.

Measured data are different and, in this respect, the rest of this chapter focuses on that.

Quality control of measured data

Real-time acquisition of metocean data should include a certain amount of quality control to ensure that displayed information is as accurate as possible. However, because information is being displayed virtually instantaneously after its acquisition, not all errors or problems may be highlighted or flagged.

This means that there is a requirement for the real-time data to be recorded and thoroughly checked on a regular basis – for example, every month or quarter. Such quality control is generally carried out by the metocean contractor responsible for the data acquisition system. However, to satisfy regulators, it is becoming increasingly a requirement to be able to demonstrate an independent review of the quality-controlled data by a third party.

Quality control (QC) is therefore a combination of 'automatic QC', applied by the processing and post-processing software to the real-time data offshore, together with 'manual QC' of the recorded data, carried out by a metocean analyst onshore.

QC is in essence looking for incorrect data and specific anomalies and identifying the same with 'error flags'. The types of things that are looked for can include inter alia:

- spikes
- outliers
- stationarity (values remaining the same over time)
- rate of change (greater than expected changes between adjacent data points in time)

The manual QC check should first make an assessment of the automatic error flags and then confirm or remove the flag. It should then carry out a range of visual inspections and comparative analyses with a view to identifying any further problems with the data. Any spurious data should then be identified with a 'manual QC flag'.

It should be noted that some apparent spikes or rate-of-change anomalies may not always be what they seem. This can be checked during the manual QC by looking at multiple parameters simultaneously. For example, a rapid change in air temperature may be associated with a sharp change in wind direction.

It is also important during the manual QC check to assess data for their 'meteorological and oceanographic appropriateness' for the location and/or time period.

Another important aspect of quality control is to ensure that certain parameters are corrected in order to comply with international standards. For example, if wind data are recorded at the top of a derrick on an offshore platform, there is a need to reduce it to a value appropriate to 10 m ASL.

Along with ensuring the quality of the data it is also important to capture and document the associated metadata (i.e. pertinent descriptive information), such as contractor, time period, data return, location, water depth, sensors etc.

Data management and archiving

Once the data have been quality controlled and fully documented the next step is to archive them in a suitable database. Metocean databases are essentially of three main types:

- those held by public bodies and/or quasi-governmental departments
- those held by commercial metocean companies
- those held by OR and O&G companies

It has already been highlighted that metocean data constitute a valuable resource that needs to be managed over the lifecycle of a project. The establishment of a long-term quality-controlled database enables metocean specialists and end-users to derive and utilise operational statistics and design criteria with confidence.

It should be noted that any erroneous data flagged during QC can either be added to the database (with appropriate caveats) or deleted and removed.

Recent technological developments and refined scientific requirements have meant that acquired data volumes (whether measured or hindcast) are increasing in size. This applies particularly to some hindcast databases, which may have 40 years of data on wind, wave, current and water levels for hundreds or thousands of grid points. Terabytes of data may therefore be expected, which may have associated impacts on the technology and hardware required to store, maintain and manage it.

Chapter 7

The development of operational statistics and design criteria

This chapter is concerned with the analysis of archived metocean data to derive statistical outputs, and with how the results are presented and used.

There are two main types of information generated from the analyses:

- Information concerned with quantifying typical or average conditions (you may also see reference to 'normal' conditions). This information is referred to as operational statistics. It is primarily required to assist with quantifying downtime, budgeting, planning and operational decision-making.
- Information concerned with quantifying extreme or design conditions (you may also see references to 'abnormal' and 'survival' conditions). This information is referred to as design criteria. It is required to assist with the design of new installations and structures, and the reassessment of existing ones.

It is important to appreciate that both operational statistics and design criteria are in essence 'predictions of the future' based on the analysis of historical data. However, there is one key difference: operational statistics are generally straightforward to produce as they are derived from relatively simple statistical analysis of past data, whereas deriving design criteria generally involves more sophisticated techniques and experienced user intervention.

Although deriving design criteria is one of the most challenging activities for metocean specialists, presenting some operational statistics in a user-friendly form can also be challenging. These issues will become clearer as the chapter progresses.

As may be expected, there can be significant differences between statistical values for operational and extreme conditions. This is illustrated in Table 7.1, which presents indicative values of mean, 1% exceedance (the 99th percentile) and 100-year Hs values for four sea areas in NW Europe.

There are a couple of quick generalisations that can be drawn from Table 7.1: firstly, the 1% exceedance value is approximately 3 times the mean value, and

Table 7.1 Indicative values for four seas in NW Europe.

Region/Sea area	Mean Hs (m)	1% Exceedance Hs (m)	Typical 100-year Hs (m)
Atlantic Margin	3.0	9.0	18.0
Northern North Sea	2.5	7.5	16.0
Central North Sea	2.0	6.5	13.5
Southern North Sea	1.5	4.5	9.0

secondly, the 100-year value is approximately 6 times the mean value. These multipliers offer an approximate 'rule of thumb' for this part of the world. It should be noted, however, that these generalisations will not necessarily apply in other parts of the world, especially for some tropical regions. And, of course, there is no substitute for actual site-specific values.

Before carrying out any metocean analyses to derive either operational statistics or design criteria, there is a need to consider the following:

- Does the end-user know what they want?
- Does the metocean specialist know what is required?
- Are the input data accurate and representative both spatially and temporally?

After carrying out the analyses there is a need to consider the following:

- Are the results 'meteorologically and oceanographically sensible', realistic and 'physically possible'?
- Is any information, briefing note or report produced from the analyses:
    - appropriate to the needs of the end-user?
    - presented in a user-friendly format with supporting legend and text?
    - provided with examples of how to interpret presentations?
    - peer (or independently) reviewed and verified?

Operational statistics

Operational statistics is a catch-all term to cover analyses and presentations of metocean data that generally focus on conditions that are important for carrying out day-to-day offshore activities. For the key data types such as winds, waves and currents this usually means low-to-moderate conditions: for example, in the North Sea Hs values are generally less than 2–3 m.

Because there is potentially a wide range of offshore activities that can benefit from operational statistics, there is also a diverse set of analyses and presentations that can be produced. These can range, for example, from simple tables of mean Hs value by month through to detailed weather windows analyses.

In all cases the statistics are designed to assist with budgeting, planning and executing operational activities, which can contribute to improved safety and reduced costs. For most applications the primary use is trying to quantify weather-related downtime and workability periods.

In this section we shall address three key aspects of operational statistics:

- important issues with respect to the data used to derive the statistics
- some examples of the types of questions that may be posed by end-users
- the types of analyses that are carried out, including the presentations produced and how to interpret and/or use the information

(1) Data issues

Operational statistics should be produced from long-term data to give a better indication of typical monthly values. For example, it is better to analyse 10 different Januarys rather than 1 because any individual month is subject to variability from one year to the next. This is known as inter-annual variability.

Traditionally, operational statistics have tended to be presented as monthly averages or means based on *x* years of data. But there is rarely an 'average month' so, increasingly, other statistical indices are provided, such as different probability levels. e.g. P10 or P90 (10% or 90% probability). However, this increases the amount of information generated, and also presents challenges for end-user decision-making: for example, what level of probability to choose.

There is also often a need to review any operational statistics in conjunction with short-term weather forecasts nearer the time of the offshore activity.

(2) End-user questions requiring operational statistics

Some examples of the types of questions posed by end-users, together with the reasons they are asked, are given here.

- How often is Hs above or below 2.5 m, 3 m and 3.5 m in a particular month?

A wide range of activities may request this type of output. For example, vessels often require Hs<2.5 m in order to carry out their activities safely and effectively.

Such questions may also be further refined by conditional requirements – for example, the installation of a structure may require a combination of Hs<1 m and Tp<12 sec.

- How many weather windows of 3 days with Hs<2 m are there in a particular month?

This is required to assess the likelihood of completing a single activity or operation.

- How long will it take to complete an operation with three different Hs thresholds starting on a particular date? Or, conversely, if an operation needs to be completed by a particular date when should it start?

This is required in order to assess the likelihood of completing complex sequential or parallel activities.

(3) Types of analyses and presentations

There are three main types of analyses relevant to the production of operational statistics:

- Basic statistics
  - minima, means and maxima of parameters
- Frequency distributions
  - based on one parameter (referred to as single frequency distribution or SFD); also called univariate
  - based on two parameters (referred to as joint frequency distribution or JFD); also called bivariate

- Weather downtime analyses (based on)
  - (i) percentage exceedance
  - (ii) persistence or duration statistics
  - (iii) detailed weather window analysis

(a) Basic statistics

Basic statistics provide a useful initial overview of the conditions recorded in a database by tabulating monthly minima, means and maxima by individual month, either as an average or all of them individually.

Thus, for example, you can have an average January mean Hs and/or mean Hs values for each January. However, such statistics generally do not provide sufficient information for operational decision-making.

(b) Frequency distributions

A *single frequency distribution* (SFD) is a term used to cover a simple presentation of the number of occurrences of a parameter for set intervals from the lowest to the highest value: for example, the number of occurrences of Hs in 0.5 m intervals. The data can also be expressed as a percentage or parts per thousand. The distribution can be presented either in tabular or graphical format.

A simple example is provided in Table 7.2, where 87 Hs values are tabulated in 0.5 m intervals. There are 3 occurrences of Hs between 0 m and 0.5 m, 8 occurrences between 0.5 m and 1 m, and so forth.

The information in the SFD can also be presented in a cumulative fashion, in two ways: firstly, looking at occurrences less than a threshold Hs (Table 7.3) and, secondly, looking at occurrences greater than a threshold Hs (Table 7.4).

In Table 7.3 there are 3 values less than 0.5 m, 11 less than 1 m and so on. Conversely in Table 7.4 there are 87 values greater than 0 m, 84 greater than 0.5 m etc. Table 7.4 is in fact a preliminary step in producing percentage exceedance information (see below).

A *joint frequency distribution* (JFD) takes things a stage further by adding in a second parameter so that one can be plotted against the other – for example, Hs versus wave direction.

Table 7.2 Tabular presentation of a singular frequency distribution for Hs.

	SFD - Occurrences of Hs in 0.5 m intervals							
	3	8	15	20	17	12	8	4
Hs(m)	0–0.5	0.5–1.0	1.0–1.5	1.5–2.0	2.0–2.5	2.5–3.0	3.0–3.5	3.5–4.0

Table 7.3 Cumulative presentation – less than a threshold Hs.

	SFD - Cumulative occurrences of Hs less than							
	3	11	26	46	63	75	83	87
Hs(m)	<0.5	<1	<1.5	<2	<2.5	<3	<3.5	<4

Table 7.4 Cumulative presentation – greater than a threshold Hs.

SFD - Cumulative occurrences of Hs greater than								
	87	84	76	61	41	24	12	4
Hs(m)	>0	>0.5	>1	>1.5	>2	>2.5	>3	>3.5

As with SFDs the presentation for JFDs can be as number of occurrences, as a percentage, or as parts per thousand. It can also be presented either in tabular or graphical format. In the latter case, the presentation is usually referred to as a 'rose' plot when one of the parameters is directional data.

Table 7.5 gives an example of a simple JFD using the data presented above for the SFD (i.e. 87 values of Hs) but including wave direction as the second parameter.

Table 7.5 shows, for example, that of the 8 waves between 0.5 m and 1 m, 3 are from the north, 2 from the northeast, 1 from the west and 2 from the northwest.

It should be noted that there is a wide range of different combinations of the two parameters that may need to be considered aside from Hs versus wave direction. For example:

- Hs versus Tz; Hs versus Tp; Hs versus wind speed
- wind speed versus wind direction
- current speed versus current direction
- wind–wave misalignment (varied by Hs and wind speed)

For each one, consideration may also need to be given to monthly analyses as well as to all-year analyses. It is also important to stress here that, although JFD presentations are derived from the analysis of past data, and in this respect can be viewed as operational

Table 7.5 Example of a simple JFD with wave direction as the second parameter.

JFD – Hs versus wave direction								
Total	3	8	15	20	17	12	8	4
Direction								
N	1	3	3	3	2	2	1	2
NE		2	2	3	2	2	1	
E			2	3	2	1	1	
SE			1	1	1	1		
S			1	1	1			
SW			2	2	2	1	1	
W	1	1	2	3	3	2	2	1
NW	1	2	2	4	4	3	2	1
Hs (m)	0–0.5	0.5–1.0	1.0–1.5	1.5–2.0	2.0–2.5	2.5–3.0	2.0–3.5	3.5–4.0

statistics, some of them can also be necessary for design criteria requirements, particularly with respect to fatigue analysis: for example, Hs versus Tz (often referred to as scatter diagrams).

(c) Weather downtime analyses

(i) Percentage exceedance

In the section on frequency distributions we introduced the cumulative approach, which is the starting point for presenting percentage exceedance statistics, which can be useful in providing potential downtime information.

The simplest way to convey downtime information is by means of a monthly *percentage exceedance* table. It therefore requires that the metocean data are analysed on a monthly basis to generate in effect 12 separate cumulative SFDs to cover the complete year.

Traditional presentations give a table with average monthly exceedance values at designated intervals for the particular parameter, such as Hs. Such a table can give the average downtime/uptime as a monthly percentage of time, but not how and when it occurs. It therefore provides some initial information to yield coarse assessments of likely downtime. But, because the values are averages, there is no information on the potential inter-annual variability.

As a result of this the author further developed this presentation (in the 1990s) to also include the minimum and maximum exceedance values from the database alongside the average. The minimum value equates to the most benign month in the dataset, whereas the maximum value equates to the stormiest. An example of this presentation is provided in Table 7.6.

Table 7.7 provides a section of Table 7.6 to help show how to interpret the information. If we consider an Hs threshold of 2.5 m, the average exceedance value for April from the table is 45%. This means that in an average April Hs is above or exceeds 2.5 m for 45% of the time. Conversely it will be below 2.5 m for 55% of the time.

Such a table should be based on many years of data – say at least 10 years – to give a better representation of the average. In the above example (derived from 30 years of actual North Sea data) the ‘Min’ value for April is 9%, which represents the least stormy April, whereas the ‘Max’ value of 70% represents the stormiest. This shows the potential for considerable variation in expected downtime from one year to the next.

Percentage exceedance information can also be presented graphically. Although such a presentation gives a good visual appreciation of the data, there will still be a requirement to scale off the required values from the lines.

Figure 7.1 shows monthly Hs exceedance lines for a coastal area in the North Sea. For example, a 2m exceedance value is around 20% in January but only around 5% in July.

(ii) Persistence or duration statistics

Persistence analysis takes operational statistics a stage further from percentage exceedance. It is a technique to identify weather windows of particular durations for nominated thresholds of a parameter such as Hs: for example, the number of occurrences of Hs<2 m lasting 1, 2 and 3 days by month.

Table 7.6 Monthly percentage exceedance table for Hs.

Hs ≥	Jan			Feb			Mar			Apr			May			Jun			Jul			Aug			Sep			Oct			Nov			Dec			All Year
[m]	Min	Av.	Max	Min	Av.	Max	Min	Av.	Max	Min	Av.	Max	Min	Av.	Max	Min	Av.	Max	Min	Av.	Max	Min	Av.	Max	Min	Av.	Max	Min	Av.	Max	Min	Av.	Max	Min	Av.	Max	Av.
0.0	100	100	100	100	100	100	100	100	100	100	100	100	100	100	100	100	100	100	100	100	100	100	100	100	100	100	100	100	100	100	100	100	100	100	100	100	100
0.5	100	100	100	95	100	100	100	100	100	99	100	100	93	100	100	96	100	100	88	99	100	91	99	100	97	100	100	100	100	100	100	100	100	100	100	100	100
1.0	97	100	100	91	99	100	93	99	100	68	96	100	69	90	100	64	86	100	51	80	97	62	83	99	84	97	100	95	99	100	97	100	100	93	99	100	94
1.5	86	98	100	74	95	100	74	93	100	26	81	98	36	64	97	27	55	88	20	45	67	32	55	74	46	81	100	78	92	100	87	96	100	74	96	100	79
2.0	70	92	100	43	86	100	55	83	100	15	62	88	11	38	66	12	30	60	2	22	45	6	30	51	15	59	85	58	78	100	60	87	100	57	89	100	63
2.5	54	83	100	21	75	100	37	70	99	9	45	70	4	22	45	5	15	30	0	9	25	2	14	34	10	41	64	34	61	89	33	73	98	43	78	98	49
3.0	40	70	96	8	61	94	27	56	95	2	30	57	0	12	28	0	7	19	0	4	15	0	6	19	3	27	48	20	45	77	21	56	90	28	65	89	36
3.5	26	57	88	3	49	86	19	42	84	0	19	40	0	7	19	0	3	13	0	2	7	0	3	13	1	17	37	9	31	64	15	42	71	17	50	77	27
4.0	13	46	80	1	37	78	11	31	69	0	12	28	0	3	15	0	2	10	0	1	5	0	1	9	0	10	30	4	21	51	6	31	60	9	38	63	19
4.5	8	36	72	0	27	67	7	22	52	0	7	18	0	2	12	0	1	5	0	0	3	0	1	5	0	6	20	1	14	37	2	22	47	5	27	51	14
5.0	4	27	63	0	20	54	4	15	41	0	4	13	0	1	9	0	0	4	0	0	3	0	0	3	0	4	13	0	9	25	0	16	36	3	18	41	9
5.5	1	19	52	0	14	41	2	10	31	0	2	10	0	0	7	0	0	3	0	0	3	0	0	2	0	2	8	0	6	19	0	10	28	1	12	27	6
6.0	0	13	40	0	9	31	0	6	20	0	2	8	0	0	5	0	0	3	0	0	2	0	0	2	0	1	5	0	3	13	0	7	22	0	8	17	4
6.5	0	9	32	0	6	22	0	4	15	0	1	6	0	0	3	0	0	2	0	0	1	0	0	2	0	1	4	0	2	9	0	5	15	0	5	13	3
7.0	0	6	24	0	4	17	0	3	11	0	1	5	0	0	2	0	0	2	0	0	1	0	0	1	0	0	3	0	1	7	0	3	11	0	3	9	2
7.5	0	4	20	0	2	15	0	2	6	0	0	3	0	0	2	0	0	1				0	0	1	0	0	3	0	1	7	0	2	8	0	2	6	1
8.0	0	3	16	0	1	11	0	1	5	0	0	3	0	0	2	0	0	1				0	0	1	0	0	1	0	0	6	0	1	6	0	1	5	1
8.5	0	2	13	0	1	9	0	1	4	0	0	3	0	0	2	0	0	1				0	0	1	0	0	1	0	0	6	0	1	4	0	1	5	0
9.0	0	1	9	0	0	6	0	0	3	0	0	2				0	0	1				0	0	1	0	0	1	0	0	5	0	0	3	0	0	4	0
9.5	0	1	7	0	0	5	0	0	3	0	0	1				0	0	1				0	0	1				0	0	5	0	0	3	0	0	3	0
10.0	0	1	5	0	0	4	0	0	3	0	0	1																0	0	4	0	0	2	0	0	3	0
10.5	0	0	3	0	0	3	0	0	3																			0	0	3	0	0	1	0	0	3	0
11.0	0	0	2	0	0	2	0	0	3																			0	0	3	0	0	1	0	0	3	0
11.5	0	0	1	0	0	1	0	0	1																			0	0	1	0	0	1	0	0	2	0
12.0	0	0	1	0	0	1	0	0	1																			0	0	1				0	0	2	0
12.5	0	0	1																									0	0	1				0	0	1	0
13.0	0	0	1																															0	0	1	0
13.5	0	0	1																															0	0	1	0
14.0	0	0	1																															0	0	1	0
14.5	0	0	1																																		0

Table 7.7 A section of the monthly percentage exceedence table for Hs.

Hs ≥ [m]	Jan			Feb			Mar			Apr		
	Min	**Av.**	Max	Min	**Av.**	Max	Min	**Av.**	Max	Min	**Av.**	Max
0.0	100	**100**	100	100	**100**	100	100	**100**	100	100	**100**	100
0.5	100	**100**	100	95	**100**	100	100	**100**	100	99	**100**	100
1.0	97	**100**	100	91	**99**	100	93	**99**	100	68	**96**	100
1.5	86	**98**	100	74	**95**	100	74	**93**	100	26	**81**	98
2.0	70	**92**	100	43	**86**	100	55	**83**	100	15	**62**	88
2.5	54	**83**	100	21	**75**	100	37	**70**	99	9	**45**	70
3.0	40	**70**	96	8	**61**	94	27	**56**	95	2	**30**	57
3.5	26	**57**	88	3	**49**	86	19	**42**	84	0	**19**	40
4.0	13	**46**	80	1	**37**	78	11	**31**	69	0	**12**	28
4.5	8	**36**	72	0	**27**	67	7	**22**	52	0	**7**	18

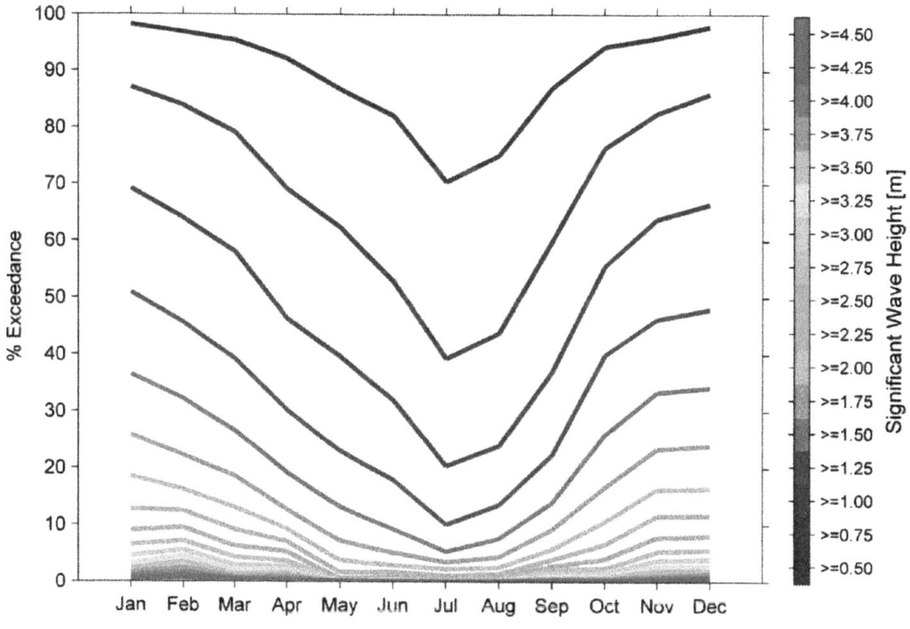

Figure 7.1 A graphical presentation of percentage exceedance.

Historically results have been presented for the average monthly condition, but more recently other probability levels such as P10, P25, P75 and P90 have also been considered. A simple example is presented in Table 7.8, which gives the average condition using actual North Sea data.

Table 7.8 Average number of windows for various durations. Hourly mean wind speed ≤10.0 m/s.

	Window duration (hours)									Wind speed <10m/s		Wind speed >10m/s	
Month	12	24	36	48	60	72	96	120	144	Av	Max	Av	Max
January	20.8	7.6	3.8	2.3	1.4	1.0	0.5	0.3	0.2	18.0	88.0	23.3	101.2
February	22.3	8.9	4.7	3.0	1.8	1.3	0.7	0.3	0.2	22.7	101.0	24.3	98.9
March	26.7	10.4	5.7	3.4	2.3	1.7	0.8	0.4	0.2	24.0	112.4	20.6	78.6
April	35.0	15.0	8.6	5.5	3.7	2.9	1.6	1.0	0.7	35.6	177.1	16.9	55.9
May	42.4	18.8	11.4	7.6	5.5	3.9	2.4	1.6	0.9	41.7	181.5	14.5	48.3
June	43.6	20.0	12.2	8.2	6.0	4.7	3.0	2.0	1.5	51.8	244.9	12.7	40.1
July	48.9	22.4	13.5	9.5	7.2	5.4	3.6	2.5	1.9	58.6	257.6	11.2	32.0
August	45.9	20.7	12.8	8.5	6.3	4.6	3.0	2.1	1.3	47.9	210.6	12.2	35.3
September	34.2	14.4	8.2	5.2	3.8	2.8	1.5	1.0	0.8	32.1	158.3	17.3	59.0
October	29.2	11.7	6.3	3.8	2.4	1.7	0.8	0.6	0.3	25.4	117.3	20.4	81.8
November	25.9	10.2	5.2	3.2	2.1	1.5	0.7	0.3	0.2	22.8	99.9	21.2	81.5
December	23.8	9.1	4.7	2.9	1.6	1.2	0.8	0.4	0.3	21.4	99.8	23.7	85.2

If we take the month of January in Table 7.8, the following information can be drawn:

- Values in the window duration columns 12 to 144 hours are average durations for wind speeds <10m/s. Thus, in an average January there are 20.8 occurrences of weather windows lasting 12 hours, 2.3 occurrences of weather windows lasting 48 hours, and so on.
- Values in the next 2 columns are average and maximum durations when the wind speed is <10m/s. Thus, in an average January when the wind speed is <10m/s it remains so on average for 18.0 hours; and the maximum time it has been continuously <10m/s is 88.0 hours.
- Values in the final 2 columns are average and maximum values when the wind speed is >10m/s. thus in an average January when the wind speed is >10m/s it remains so on average for 23.3 hours; and the maximum time it has been continuously >10m/s is 101.2 hours.

It should be noted that persistence analyses can generate considerable numbers of different outputs. For example, Table 7.8 above can be produced for wind speeds >10m/s plus equivalent tables for other speeds and other probability levels. In certain cases this can lead to the production of literally hundreds of tables. This is one of the reasons why it is important for there to be dialogue between metocean specialists and end-users to ensure that a sensible (and relevant) amount of output is produced.

(III) DETAILED WEATHER WINDOWS ANALYSIS

Standard persistence analysis is primarily geared to one operation with one threshold condition. Some offshore activities, however, are quite complex in that there may consist of a range of activities happening sequentially or in parallel that require different thresholds. Under such circumstances a more sophisticated analytical technique is required, generally referred to as detailed weather windows analysis.

An example of a multi-threshold series of operations could be as follows:

- Operation 1 – requires Hs<1.5 m for 1 day
- Operation 2 – requires Hs<1 m for 0.5 day
- Operation 3 – requires Hs<1.5 m and Tp<12 s for 2 days

In some situations all operations in a sequence must take place without interruption; in others it may be possible to complete Operation 1 and wait for a second weather window for Operations 2 and 3. The software for these analyses can cater for both situations, as well as for instances where a specific operation cannot be stopped once it has started.

There are also different permutations for end results, depending on user requirements. For example:

- having a fixed start date for the operations and wanting to know when activities will be completed
- having a fixed end date and wanting to know when to start
- having potentially different start dates and wanting to know how long it will take for each, plus the waiting time before activities can start

As with persistence analysis, results can be presented for a range of probability levels. Again, this tends to generate a lot of output, which can pose problems for those producing the information, i.e. how to present it in as user-friendly format as possible. It also poses challenges for end-users in how to interpret the information and what probability level to use.

Although weather downtime analyses can provide useful information, valuable insight can be obtained by carrying out a 'helicopter assessment' of the long-term database. The use of the term 'helicopter assessment' refers to viewing all the data at once. This can be achieved by producing what the author has termed the 'Supermarket barcode overview'.

Figure 7.2 provides an example of this by looking at all the occurrences of Hs<1.5 m for 1 day between 1992 and 2008 for an offshore location. Black bars indicate when

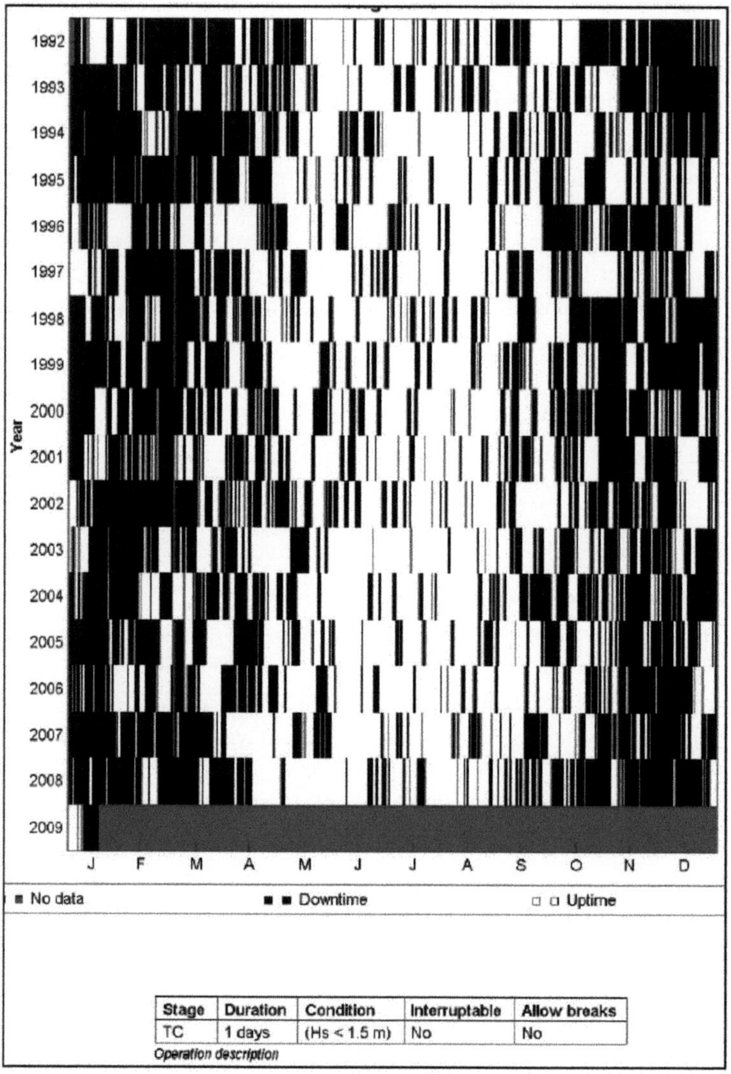

Figure 7.2 Barcode chart.

this requirement was not met. Conversely, white shows when work could have taken place.

Figure 7.2 clearly shows the expected seasonal bias, i.e. more opportunities to work in the summer months. But it also shows the inter-annual variability of particular months. For example, contrast October 1997 with October 1998 – it would clearly have been possible to carry out operations throughout October 1997 whereas there were no opportunities in October 1998. This is an excellent way of viewing when activities could have taken place in the past.

Design criteria

The previous section has considered various operational statistics that focus primarily on low-to-moderate metocean conditions. In this section we shall look at the derivation of extreme or design statistics that quantify the most severe metocean conditions.

All offshore installations need to be designed and built to withstand the extreme conditions expected in their operating area. For example, if we consider waves, which in many cases exert a dominant effect on design issues, there are three key pieces of metocean information needed for design:

- The height and period of waves. These impact the speed and depth penetration of water particle velocities (associated with the circular motion of waves as they pass through or impact a structure), which in turn exert drag forces on elements of the structure.
- The total vertical distance between the sea bed and the highest wave crest. This is important because very high overturning moments (otm), forces that cause a structure to become unstable and turn over, are induced if the wave crest impacts the upper part of the structure rather than the monopile or the lattice work of the legs.
- The distribution of individual wave heights and their associated periods. These are needed in order to calculate the expected fatigue life.

Optimising metocean design criteria is important to ensure a safe yet cost-effective design. In this respect there is a need to avoid under-design, which could have an adverse impact on safety, people, equipment and the installation. Conversely, there is a need to avoid over-design, which can lead to increased cost.

It should be stressed that, in providing design criteria, we are not designing for the worst foreseeable condition but, rather, taking a statistical or probability-based approach related to an *n*-year return period.

For most requirements the default return period for criteria is 50 years for OR and 100 years for O&G. It needs to be stressed that this does not equate to 'happening 50 or 100 years from now' but rather to it being exceeded once on average during those 50 or 100 years.

This *n*-year return period can also be expressed in terms of probability. Thus, a 100-year value equates to a 1% probability of being exceeded in any one year, i.e. 10^{-2}. For a 50-year return period that probability increases to 2%.

The derivation and selection of design criteria should follow a recognised process. The process follows sequential steps that generate some of the criteria in the first step, then others in the second that depend on those generated in the first, and so on.

These steps give rise to what are termed primary, secondary and tertiary criteria. It should be made clear that these designations do not imply level of importance: they are simply a stage in the process.

Whilst the main focus of this chapter is concerned with deriving 50-year or 100-year values, it should be appreciated that other return period values are sometimes needed, whether they be shorter (say 1-year and 10-year) or longer (10,000-year) as in the case of extreme water levels (EWL).

In order to differentiate between 50 or 100-year and 10,000-year return periods they are often classified as extreme and abnormal events respectively. Extreme equates to the design condition whereas abnormal equates to a survivability check.

A final point to make in this introduction concerns the issue of fatigue life, mentioned in the third bullet point above as one of the three key pieces of metocean information needed for design. The requirements for design criteria do not just involve the derivation of specific numbers or values: they also involve the provision of various analytical presentations, usually in the form of a JFD of two parameters – for example, wave height versus wave period, or wave direction versus wind direction.

The next section will detail the process of deriving the full range of criteria and the analyses required to achieve this.

How to derive design criteria

As mentioned previously, deriving 50-year or 100-year design criteria can be challenging. In order to derive realistic criteria there is a need for the following:

- knowledge of the area or location
- familiarity with legislation and standards
- a quality and representative long-term database
- appropriate analytical tools and software
- competent metocean contractors and consultants
- dialogue between the metocean provider and the end-user
- independent verification of criteria

It should also be noted that whilst the derivation of design criteria follows a technical process, there is also an associated administration process that should be adhered to. The key steps in the technical process are outlined first, followed by the administration process.

(1) Key steps in the technical process

The key steps in the technical process are:

- Identify appropriate datasets, which should ideally be both measured and hindcast data
- Prepare and format the datasets for analysis focusing on Hs, 1-hour mean wind speed (10 m ASL) and depth-mean or current sensor depth total speed; these are referred to as *primary parameters*

- Use extreme value analysis (EVA) to extrapolate data to *n*-year return periods, focusing on 50-year or 100-year *all-year omni-directional values*
- Ensure that a number of different EVA runs are carried out: for example, by varying input data and extrapolation techniques
- From these various extreme values establish the final 50-year or 100-year design criteria for the primary parameters, focusing on the best fits
- Derive the 50-year or 100-year design criteria for those parameters dependent on the primary ones, such as 3-sec wind gust, wind profile, Hmax, Tz and current profile; these may be called *secondary parameters*
- Derive the 50-year or 100-year design criteria for those parameters dependent on the secondary ones, such as Tp and extreme water level (EWL), noting that EWL requires 10,000-year criteria; these may be called *tertiary parameters*
- Derive other return periods, such as 1-year and 10-year, seasonal and monthly criteria, directional criteria, and other parameters such as extreme air and sea temperature
- Produce a range of 2-parameter JFDs primarily for fatigue assessments and specific structure/installation requirements
- Prepare design criteria tables and figures, and produce a report

These key steps are expanded upon in subsequent sections.

It is important to note again that the terms primary, secondary and tertiary do not imply the relative importance of the parameters. The terminology is merely used to illustrate the requirements and sequence of the technical process. For example, although EWL is classified as tertiary in the process, it is one of the most important parameters.

(2) Key steps in the administration process

Whilst the technical process is clearly important, it is also essential to ensure that the associated administration process is adhered to, as this can help contribute to an optimal set of design criteria and a quicker turnaround time for the final report. A recommended approach for this would be:

- an OR/O&G company requires a metocean design report for a new or existing installation
- if the company has no in-house metocean expertise, they should consider using an independent metocean consultant to assist with preparation of the scope of work, the specifications, and the identification of competent contractors
- the detailed specification should take account of what the design criteria are required for – for example, a wind turbine, a floating production, storage and off-take (FPSO) etc.
- a tender document is sent out to competent metocean contractors
- tenders are evaluated technically and financially, and a contract is awarded
- the metocean contractor may find it useful to consider utilising an independent consultant to ensure their analyses and recommendations are sensible. As part of this there should be regular dialogue between the OR/O&G company (or its independent consultant) and the metocean contractor to ensure a quality product

- the contractor issues a draft report for review and comment by the OR/O&G company or their independent consultant
- the final report is issued and the design criteria are accepted

Unfortunately these processes are not always adhered to, which can result in design criteria being issued that are erroneous for the location or inappropriate for a particular installation type. For example, if there is no checking of the contractor work and results, the design wave heights may be too high or too low, or the wave periods may be in error by several seconds. In some circumstances this could lead to project delays or even impact the ability to proceed.

Extreme value analysis

The key analytical part of the technical process is the use of extreme value analysis (EVA) to estimate conditions that occur at very low probability levels which equate to *n*-year return periods. In essence this means effectively going from the known (for example 30 years of data) to the unknown (50-year or 100-year extremes).

In order to carry out the EVA there is a need to firstly prepare the source data to be used, and secondly to consider appropriate probability levels, functions and fitting techniques

(1) Source data

The metocean data (for example, time series of Hs or 1-hour mean wind speed) for input into EVA is generally prepared in two main ways:

- Using all the data from the lowest to the highest values and placing them in equal ascending intervals, known as 'binning'. For example, Hs data may be binned in 0.5 m intervals; thus, there will be *x* occurrences between 0 m and 0.5 m, *y* occurrences between 0.5 m and 1.0 m, and so on. This type of approach is referred to as cumulative frequency distribution (CFD) and is analogous to the development of SFDs for operational statistics.
- Using a subset of the data by focusing exclusively on peak values above a nominated threshold. For example, Hs values >7 m. This type of approach is referred to as peaks over threshold (POT).

(2) Probability levels and fits

In both cases the source data are plotted as data points on logarithmic axes in terms of their 'probability of non-exceedance', the value of which depends on the number of observations that occurs over the return period in question:

$$P(x) = 1 - 1/(365.25mRP)$$

where

$P(x)$ = probability of non-exceedance
m = number of observations in a day
RP = return period (years)

Thus, the probability level is a function of the measurements within the source data, as illustrated by the following example:

- If you measure the wave height for 20 mins at 3-hour intervals for 100 years you will get 292,200 Hs measurements (365.25*8*100).
- The chance of any one measurement being the highest is 1/292,200 = 0.00000342: this is the probability of exceedance.
- Conversely, the chance that any one measurement is less than the highest is 1–(1/292,200) = 0.99999658: this is the probability of non-exceedance.

The plotted data points are fitted to an appropriate function – for example, Weibull – and extended beyond the available data to *n*-year return periods at appropriate probability levels. Thus, as mentioned above, the probability level of 0.99999658 is equivalent to a 100-year return period for a 3-hour dataset.

Figure 7.3 shows a typical graphical output from an EVA analysis. It shows an extrapolation of 34 years Hs data with points in blue plotted in terms of non-exceedance; red crosses show the intersections for return periods from 1 year to 100 years; and a best-fit line through the data points is shown in black.

In this particular example the 100-year extreme value is 7.49 m based on:

- the top 70 peak values of Hs in the dataset
- a Weibull 3 distribution and
- a fit method known as maximum likelihood.

Data Source	
Dataset Name	T10
Dataset Duration	34.00 Yr
Sample Interval	1.00 Hr
Percent Valid	100.0
Threshold	4.31
Reset Duration	72 Hr
Number Of Peaks	70
Storms Per Year	2.06
Model Fitting	
Model	WEIB3
Fit Method	MLE
R^2	0.9729
Fitted Weibull Distribution	
Location	4.3106
Scale	0.6948
Shape	1.1000

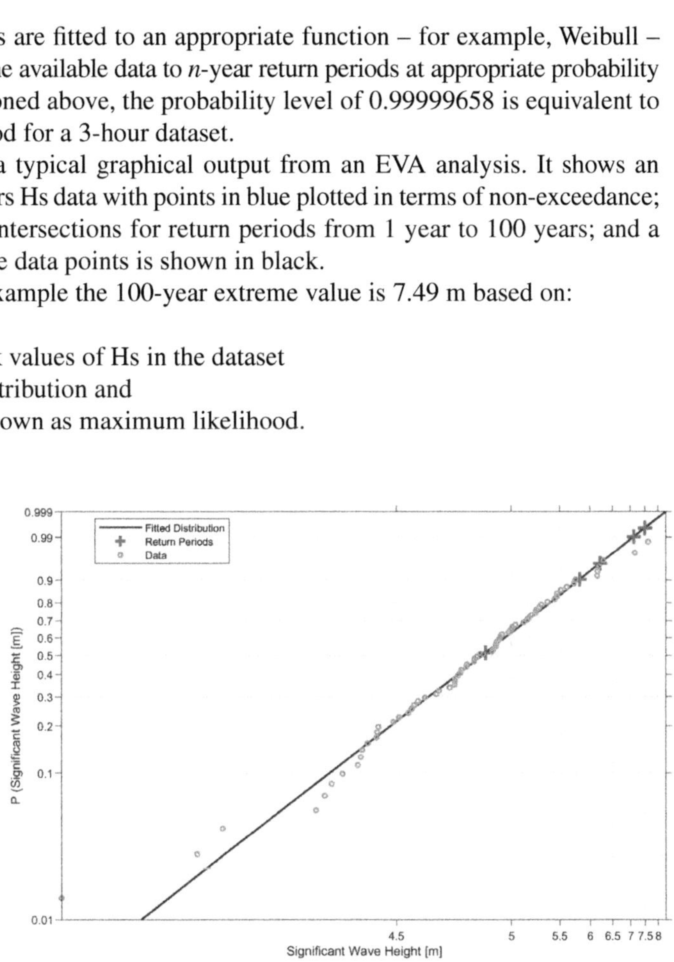

Return Period	Probability	Extreme Value
1	0.5143052704	4.83
5	0.9028610541	5.81
10	0.951430527	6.21
50	0.9902861054	7.11
100	0.9951430527	7.49

Figure 7.3 An EVA graphical output (courtesy of Metoceanworks Ltd).

It should be noted that the best-fit line in this example will change depending on the number of peaks selected, the choice of distribution, and the fitting technique. Each change of fit will result in a different set of extreme values. Further information on EVA is provided in Appendix 4.

From extreme values to design criteria

The 50-year or 100-year extreme values generated from the multiple EVA runs pose a key challenge: How do you go from a number of different extreme values of, say, Hs or 1-hour mean wind speed, to the final design value?

As a pre-cursor to this question, it is useful to elaborate further on the reasons why you can get so much variability in extreme values. The main reasons for this relate to (1) the input data used (2) how the EVA is carried out, and (3) what company/person is doing the work.

(1) Input data

Variability as a result of input data can be due to:

- type of data used (measured versus hindcast)
- quantity of data used, the overall data return and its quality
- time period analysed and the representativeness of the data
- sampling duration and recording interval

(2) Extreme value analysis

Variability as a result of EVA can be due to:

- type of function or distribution used
- changes in the position of the best-fit line due to, for example,
    - proportion of data used (for example, the top 20% versus 95%)
    - thresholds selected (for example Hs>7 m versus >8 m etc.)

(3) Company/people

Variability as a result of the company/people can be due to:

- companies having their own specific preferences for EVA
- the subjectivity of individuals

Selection of criteria

Each set of 50-year or 100-year extreme values should also have an assessment of the quality of fit. Normally only the best fits are considered. The selection of final criteria from the best fits should be based on:

- a consideration of the input data, EVA, and who carried out the analyses
- knowledge of the specific offshore location/area together with expectations for what the criteria may be. Maps of indicative criteria are available, such as the North Sea example shown in Figure 7.4
- metocean expertise and experience, often referred to as engineering judgement

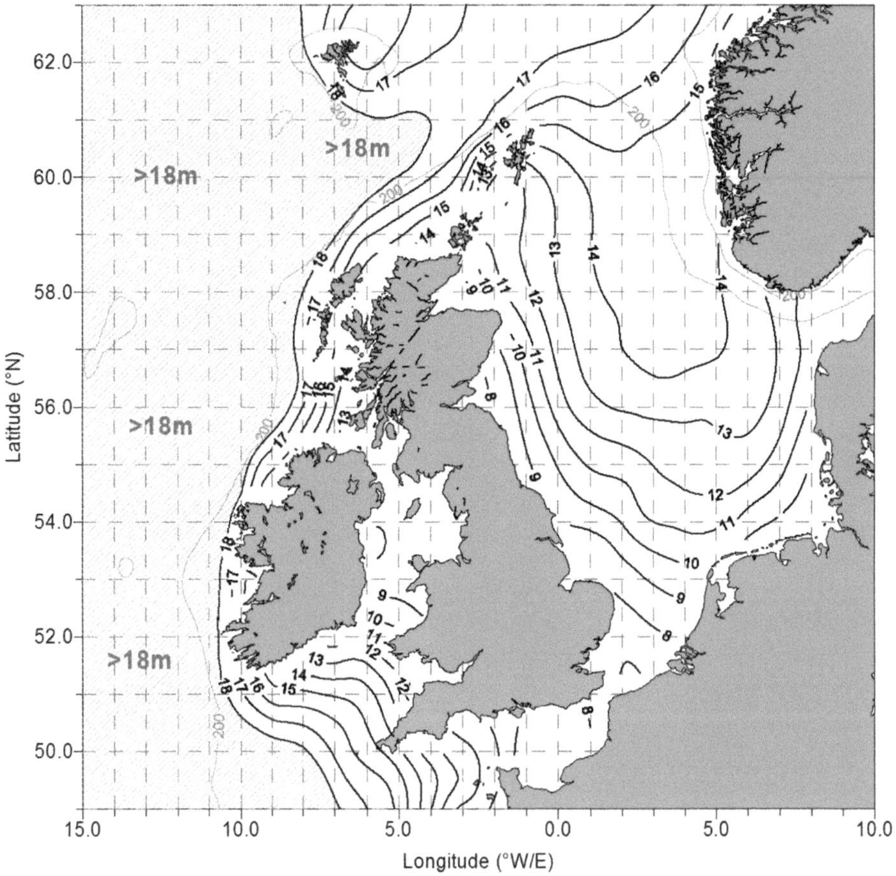

Figure 7.4 Indicative 100-year Hs design values around the UK (from HSE document RR392).

For a development in the Atlantic Margin, two major O&G companies produced more than 240 different extreme values of Hs. The results ranged from 12 m to 20 m, which illustrates both the variability of extreme values and the potential danger of just doing one EVA and accepting that as the answer. From those 240 values, a final Hs 100-year design value was selected based on the 3 bullet points above.

Reliance on one EVA run is not the only potential concern: problems can also arise with dependency on computer output. A computer-derived 'best-fit' through the data may appear to be 'statistically correct' but may not reflect reality, particularly if the data points do not enable a sensible (visual) fit. For example, a 'curved' set of data points may encourage a straight best-fit line, which could result in a value over- or under-estimating the actual 100-year value.

All this means that there is the potential for the optimal selection of design criteria for the same location depending on the input data, the contractor used and the approach taken by them. *The use of and input from experienced metocean specialists is therefore essential to ensure the selection of optimal criteria.*

Design criteria for primary parameters

The previous sections have discussed the use of EVA to derive extreme values and design criteria focusing on winds and waves. The derivation of extreme values and design criteria for currents can follow a similar approach by using EVA to extrapolate the total or observed current. However, although this method can provide reasonable results, it is often better to consider analysing the two main components that make up the total current, namely the tidal and residual components, separately.

When considering the separate analysis of the tidal and residual components, typically one should use an appropriate tidal current value, such as the mean spring tidal current (MSTC), and then combine it with the 50-year or 100-year return period value for the residual current derived from EVA.

These two approaches have been the standard in areas such as the North Sea where the current regime is relatively straightforward and dominated by tidal currents and storm surges.

However, in some parts of the world, particularly in deeper oceanic waters, characterised by a wide range of metocean processes and complex current regimes, extrapolating the residual current on its own can sometimes be insufficient as there may be multiple residual components.

Extrapolation of the total current is one approach and is still used by some people today, particularly where one process dominates; and, of course, it is relatively straightforward to do. But for many locations the key issue is how to combine the different residual components sensibly, and not assume a worst case scenario for each, to arrive at a realistic design current profile.

This means that deriving extreme currents and final design criteria does not necessarily follow a set pattern and very much depends exactly where you are in the world. These issues are discussed further in the section on current profiles later in this chapter.

Returning to the use of EVA and the selection of final criteria, we should now be in a position to have 50-year or 100-year return period values for the three key primary parameters, namely:

- significant wave height (Hs)
- 1-hour mean wind speed
- current speed (depth-mean for hindcast data and/or sensor depths for measured data)

At this stage these values are classified as:

- all-year: i.e. applicable for January–December
- omni-directional: i.e. applicable to all directions
- independent: i.e. assuming they all occur at the same time

We can now move on to discussing the development of criteria for the secondary and tertiary parameters.

Design criteria for secondary and tertiary parameters

This section is concerned with issues and derivation techniques for the parameters dependent on the primary ones. It focuses on the following:

- wind speeds (for a range of averaging periods and heights)
- maximum wave height (Hmax)
- wave periods
- current profiles
- extreme water levels (which includes crest elevation)

(1) Design parameters: wind speeds (averaging periods and profile)

The primary parameter of 1-hour mean wind speed at 10 m ASL is the standard reference value from which a range of other values is required because wind speed varies with averaging period and elevation above sea level. Averaging periods range from 24 hours to 3 seconds; elevations range from +10 m to +150 m ASL.

With respect to the averaging period, wind speed increases as the averaging period decreases; for elevations, wind speed increases with height. Thus for example, a 1-minute mean wind speed will be greater than a 1-hour value and a wind speed at 100 m ASL will be higher than at 10 m ASL.

It should be noted that most wind sensors used offshore by O&G are located at the top of a platform derrick, typically 50–100 m ASL. Measured wind speeds and directions are displayed for that elevation. At such elevations speeds are typically some 20% higher than those near the sea surface.

If there is a requirement for wind speeds at levels between 10 m and 100 m, the Norwegian Petroleum Directorate (NPD) recommendations are a useful reference point.

Figure 7.5 shows the wind profiles for six different averaging periods based on a 1-hour mean wind speed of 35 m/s at +10 m ASL.

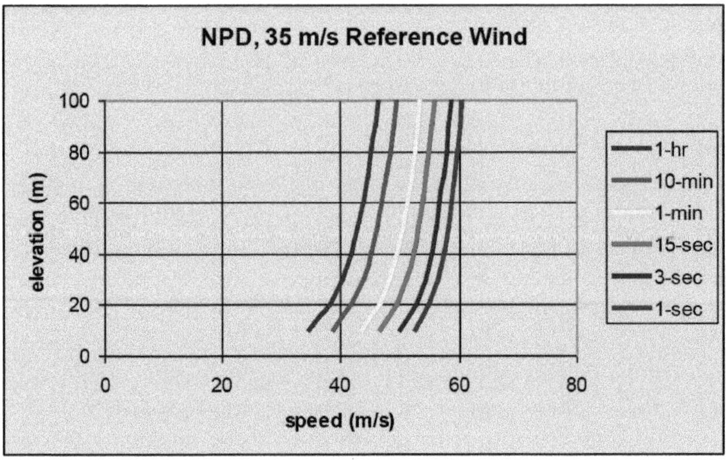

Figure 7.5 NPD wind profiles.

As noted in Chapter 4, OR may install a met mast with wind sensors at different levels. In such cases profile data should be readily available and can also be checked against existing standards such as those of the NPD.

Further information can be found in *ISO 19901-1*, which gives specific formulae for deriving both averaging period and profile values.

(2) Design parameters: maximum wave height Hmax

Terminology issues associated with Hmax were described in Chapter 2. This section elaborates on this by providing a methodology for its calculation. Historically the calculation of Hmax followed a simplistic approach as a ratio related to Hs. As a result of work carried out in the early 2000s, a new method for the derivation of Hmax was recommended and this approach has been included in *ISO 19901-1*. This effectively replaced the so-called traditional method.

(a) Historical method for Hmax calculation. This method is based on the concept of a 'design sea state' in that it assumes that the 50-year or 100-year Hmax occurs at the same time as the 50-year or 100-year Hs. It also assumes a Rayleigh distribution of individual waves in this sea state.

As a result, the ratio Hmax/Hs for this method is typically *c.* 1.85, and in simple terms this ratio is usually applied to the 50-year or 100-year Hs to determine the associated Hmax.

This method for deriving Hmax has been around since the 1970s and, although effectively superseded, is nevertheless still used by some practitioners.

(b) ISO 19901-1 method for Hmax calculation. The assumption of the 'design sea state' is not really valid because it is possible for the 50-year or 100-year Hmax to occur in other sea states: for example, the associated Hs may only be a 10-year return period event. This 'variability' of the Hmax is referred to as short-term variability, and has led to what is known as a 'storms-based' approach rather than a 'design sea state' approach.

The method used should account for the long-term uncertainty in the severity of the environment and the short-term uncertainty in the severity of the maximum wave of a given sea state or storm.

Statistically correct methods are based on storms obtained from a time series of significant wave height (Hs) by breaking it into events that have a peak Hs (Hsp) above some threshold – for example, 7 m.

The long-term uncertainty in the severity of the environment is treated using the probability distribution of the severity of the storm, measured either in terms of Hsp or the most probable maximum value of the individual waves in the storm (Hmp).

The uncertainty in the height of the maximum wave of any storm is estimated as a probability distribution conditional on Hsp or Hmp. Convolution of the two distributions gives the distribution for any random storm and, thereby, the complete long-term distribution for the heights of individual waves. Further details are provided in *ISO 19901-1*.

However, it is interesting to note that when the ISO-compliant 100-year Hmax is compared to the 100-year Hs, it generally results in Hmax/Hs ratios similar to those arrived at by the traditional method. This is because, despite the new method being more rational and more statistically sound, the traditional approach has two errors that tend to cancel each other out. In this respect it neglects short-term variability, which tends to under-estimate, whereas the use of the Rayleigh distribution tends to over-estimate.

(3) Design parameters: wave periods

There are three main wave period parameters used in design:

- mean wave period (Tz)
- peak wave period (Tp)
- associated wave period (Tass or Tmax)

The mean wave period (Tz) is associated with Hs and is related to wave steepness ($2\pi Hs/gTz^2$). Although there can be considerable variation in steepness at lower wave heights, values tend to converge and stabilise at the higher wave heights.

When providing design values it is standard practice to produce a range of Tz based on wave steepness, to provide what are termed lower, central and upper estimates. Typical steepness values in areas such as the North Sea are as follows:

Tz lower ~1:13
Tz central ~1:16
Tz upper ~1:19

In other parts of the world different ratios may apply. For example, in areas more exposed to oceanic influences and swell waves, such as the Atlantic Margin or offshore Nigeria, steepness values are more typically from 1:15 to 1:25.

The peak wave period (Tp) is associated with Hmax and equates to the maximum energy in the wave spectrum. It is generally derived by applying ratios (from analysing measured or modelled wave data) to Tz. Typical ratios for areas such as the North Sea are:

Tp lower ~1.2*Tz central
Tp central ~1.4*Tz central
Tp upper ~1.6*Tz central

As with Tz, different ratios may apply in other parts of the world.

The associated or maximum wave period (Tass or Tmax) is also associated with Hmax. It was widely used prior to the spectral analysis of wave data and the availability of Tp. It is typically 5–10% less than Tp.

(4) Design parameters: current profiles

Initial EVA runs of current data should provide extreme values and design criteria for the measurement or model depth, as discussed previously. This information has to be taken a stage further to assess and quantify the form and shape of the design current profile.

As many current models only provide depth-mean values (generally equivalent to about two-thirds the water depth) and some current sensors only provide data at fixed depths, it can sometimes be challenging to select appropriate design criteria profiles. Use of 3D models and/or ADCP sensors can be beneficial in this respect, although the latter have technical limitations near surface and near seabed.

It is possible to make some generalisations about current profiles:

- speeds decrease with depth
- wind stress can accentuate surface/near surface speeds
- speeds are retarded near the sea bed principally due to bottom friction

It is important to note, however, that these do not always apply, particularly in deeper waters off the continental shelf.

When currents were measured with several current meters on a mooring, the derivation of a design profile throughout the water column was estimated using a theoretical profile shape. Typical profiles adopted included:

- power-law profiles (such as the 1/7th) that ensured the individual 50-year or 100-year values were within the envelope of the profile
- 'slab' profiles, which assumed relatively constant values with depth
- a combination of power law and slab
- 'joining up' the 50-year or 100-year values for each current meter linearly

Power-law and slab profiles are still widely used in areas like the North Sea that are dominated by tidal currents.

The profiles above can be referred to as 'independent' in that they tend to assume that 50-year or 100-year values will occur at the same time at all depths. Such profiles can be viewed as being conservative since it is unlikely that 50-year or 100-year values will occur at the same time.

With the use of more sophisticated sensors (such as ADCPs) that provide virtually complete current profiles, plus the increasing availability of numerical model data, it has become possible to better quantify profiles, and hence derive more realistic design values.

Figure 7.6 shows various profiles from ADCP data in a water depth of *c.* 700 m presented as 50%, 10% and 1% exceedance values together with maxima. Note that these profiles are therefore based on statistics for each profile point. They are therefore examples of 'independent' profiles. Nevertheless they do illustrate some of the points made previously: for example, the general decrease of speeds with depth and the lack of data near the surface and seabed.

The above profiles present the total current. However, our increased knowledge of the complexity of currents in deeper water has enabled investigation and quantification of the different components that make up the residual part of the total current. This has resulted in a move away from so-called independent profiles to deriving one or multiple profiles that relate to the different components or modes. Such profiles may be referred to as simultaneous and they are particularly relevant in oceanic areas with complex currents, such as the Atlantic Margin. Here, the range of water masses and oceanic flows means that different types of profiles can occur depending on what processes dominate at any one time.

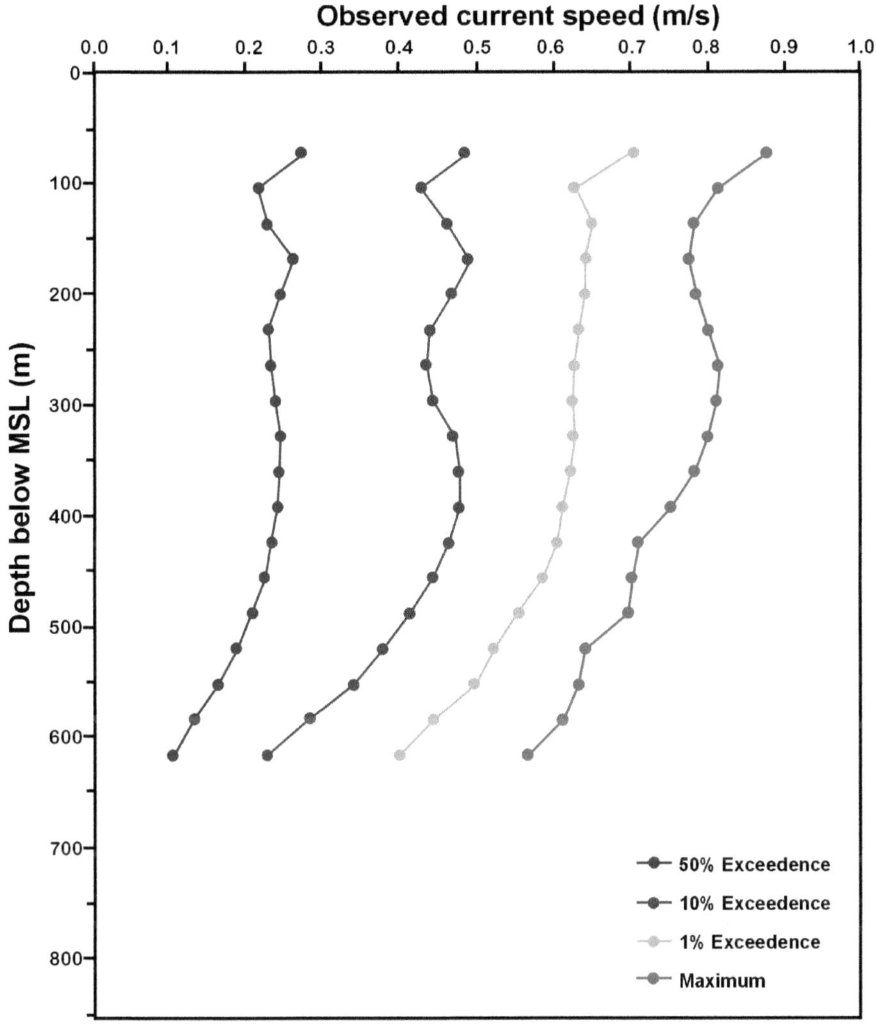

Figure 7.6 Current profiles from an ADCP.

A technique referred to as EOF analysis (empirical orthogonal functions) can be used to develop these more realistic current profiles. EOFs are used to decompose a vertical profile into a few characteristic modes that equate to the component parts of the total current.

Figure 7.7 gives an example from the Atlantic Margin and shows three profiles derived from different components or processes following EOF analysis of ADCP data. The blue line shows what might be considered a typical profile shape, with speed generally decreasing with depth, whereas the green line shows a peak speed at mid-depth.

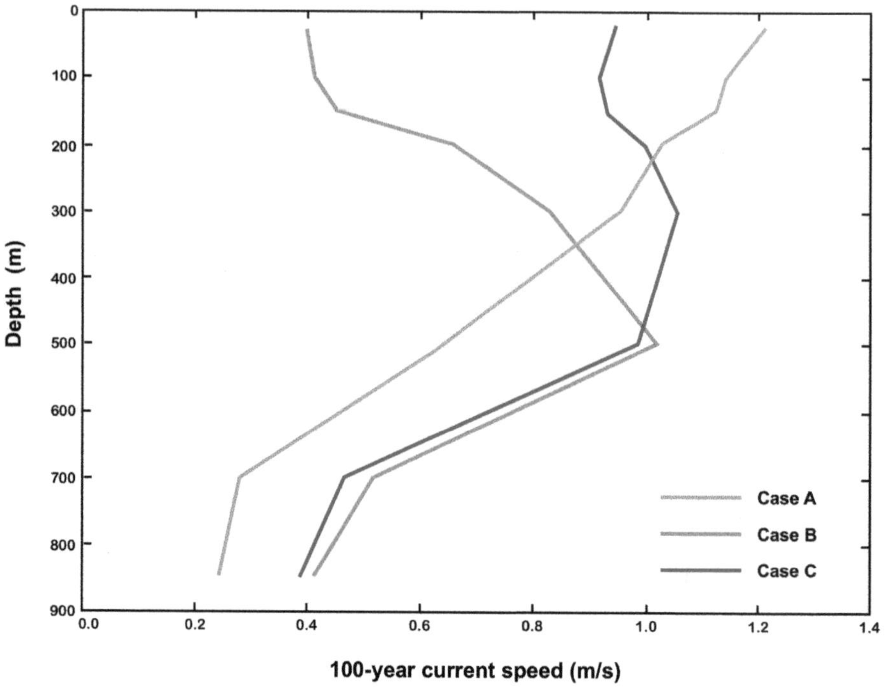

Figure 7.7 Current profiles relating to component analysis.

As with the previous discussion in the operational statistics section of this chapter on the proliferation of information generated in weather windows analyses and the challenges for end-users, multiple profile shapes also pose similar challenges in how you combine the modes and derive 50-year or 100-year values. Clearly, extrapolating the total current is easier, but it may result in higher values throughout the profile. On the other hand, deriving what is effectively 'event' 50- or 100-year values offers potential scope for reductions in the criteria.

So far we have been discussing those currents related to tidal effects and those caused by oceanic circulation, storms etc. There is also a need to consider additional horizontal motions at and near the seabed caused by waves in shallow water (typically depths <30 m) during extreme conditions. These are referred to as wave-induced currents.

Sea surface waves move forward in a circular or orbital fashion, and as water depth decreases, such waves will 'feel' the seabed, which means that the velocity of this orbital motion will be felt at the seabed. The effect of this is to generate forward momentum analogous to a current, which can be considerable and capable of moving sediment. An example of an area where such wave-induced currents are prevalent is the southern North Sea between the UK and the Netherlands.

In extreme conditions these velocities may be equal to or greater than the current speed, and if the seabed is of a sandy nature can cause problems for pipelines by removing supporting sediment.

(5) Extreme water level (EWL)

EWL is a key deliverable (particularly in the O&G sector) as it is a key input for setting the cellar deck elevation on an offshore installation. It is, therefore, one of the most important metocean parameters. The resulting value of EWL must therefore ensure that the cellar deck is not liable to wave inundation for the specified *n*-year return period.

EWL consists of three separate components of water level:

- tide
- surge
- wave crest

With respect to EWL, there are a number of challenges that providers of metocean information need to consider:

- the requirement for a 10,000-year value for manned installations rather than the standard 100-year value required for other metocean parameters, and a 1,000-year value for normally unmanned installations (NUIs)
- whether to analyse the three individual components separately and then combine in some way; or whether to directly analyse the water level data
- if analysing the components separately:
  - how to derive an appropriate value for crest elevation
  - what tidal level to use
  - what return period to use for surge to associate with the crest
- ensuring that the resulting EWL value is related to an appropriate datum (see following section)

It should be noted at this point that there is sometimes a requirement for an additional water level – namely, the still water level (SWL) – which is a combination of tide and surge only. This is normally a relatively straightforward calculation usually obtained by combining a suitable tidal level, such as MHWS, with an *n*-year return period surge.

EWL: datum

As with all requirements for information relating to data that have a vertical dimension, it is important to know to what level the EWL relates with respect to a reference position in the water column. Such a reference position is known as a datum.

You may sometimes see reference to a chart datum, which is the level of water that depths are referenced to on a nautical chart or survey map. A chart datum is generally a tidal datum, typically lowest astronomical tide (LAT), although mean sea level (MSL) may also be used.

It is therefore important to ensure that, when calculating the EWL, a decision is made as to which datum to use, and to quote it in any presentational material. Whatever datum is used it is also vital that accurate bathymetric survey data have been used to derive the LAT water depth in the first place.

Evolution of EWL methods, techniques and requirements

There have been a number of changes in methods, techniques and requirements associated with the calculation of EWL since the 1970s and these have had, and continue to have, major impacts for offshore installations. These evolutionary changes are now summarised. Although the focus is on North Sea practices, the information is broadly applicable globally.

(a) EWL: 1970s to mid-1990s

During this period, North Sea installations were designed using a formulation in the In simple terms this formulation was:

$$EWL = 50yr\ Hs \times 1.03 + Tide\ (MHWS) + 50yr\ surge$$

Once this value of EWL had been derived, structural engineers would increase the EWL slightly to allow for water depth uncertainty and any potential platform settlement. As an additional safety factor the deck elevation was then set to be at least 1.5 m above the increased EWL. This 1.5 m became known as the 'air gap', and was chosen arbitrarily in the late 1960s when O&G activity was confined to the southern North Sea, ostensibly being *c.* 10% of the typical 50-year Hmax in that area.

As activities moved northwards in the North Sea, the 1.5 m became universally applied but was clearly a smaller percentage of the Hmax: for example, it was only about 6% of the 50-year Hmax in the central North Sea.

During the 1980s the formulation changed slightly, requiring 100-year return periods following updates to the *Guidance Notes* document.

(b) EWL: issues/changes from the late 1990s onwards

Since the late 1990s there have been a number of developments and issues that have resulted in significant changes to the requirements for EWL and how it is calculated. These may be summarised as follows:

- a move from the traditional 50-year or 100-year requirement to a 10,000-year one to put the O&G industry in line with the nuclear industry where 10^{-4} is standard
- the availability of long-term hindcast metocean databases of simultaneous parameters, enabling assessment of the joint occurrence of wave and surge
- new techniques for the derivation of wave crest
- new methodologies and formulations outlined in *ISO 19901-1* and *ISO 19902*
- significant subsidence of certain platforms

The most important change relates to the move from a 100-year EWL to a 10,000-year EWL. This posed a number of serious challenges for the industry. First, there was a need to develop a pragmatic technique or method to derive such values and, second, there was concern that a 10,000-year value would undoubtedly lead to a significant increase in EWL with potential implications for existing installations. However, it should be noted that the 10,000-year EWL does not need the addition of the 1.5 m air gap.

(c) EWL: interim method for deriving 10,000-year value

The first technique for deriving a 10,000-year EWL was developed by Shell and BP who commissioned a study in the year 2000 which investigated ten different methods. This resulted in a proposed so-called 'interim method' as follows:

$$\text{EWL } 10{,}000\text{yr} = \text{Crest } 1000\text{yr} + \text{Surge } 1\text{yr} + \tfrac{3}{4}\,(\text{MHW-MSL})$$

This attempted to take account of the joint probability of occurrence of the three component parts to provide a more realistic combination of them. An update to this method was issued in 2005 and is discussed in f) below.

(d) EWL: changes to crest calculations

Studies and investigations of wave crests in the 1990's demonstrated that the empirical ratio-based methodology in the *Guidance Notes* results in a considerable underestimation of the true value.

The key findings of these studies and investigations relate to consideration of the short-term variability of individual waves in a storm, as previously discussed concerning Hmax.

The long-term distribution of extreme and abnormal crest elevations can be established from a long time series of significant wave heights (Hs) and a short-term distribution of crest elevations conditional on Hs. Further details can be found in *ISO 19901-1*.

This has typically led to an increase of circa 10% in design crests compared to the old *Guidance Notes* method for equivalent return periods. Presenting this as a ratio to 100-year Hs typically gives values in the range 1.10–1.15 compared to the 1.03 in the *Guidance Notes*.

Thus, for example, using a 100-year Hs of 15 m gives a crest of 15.45 m based on the *Guidance Notes* methodology compared to a value of the order of 17 m for an ISO-compliant 100-year crest. Equivalent ratios for the 10,000-year situation are typically 100-year Hs x 1.4–1.5, giving a crest in excess of 21 m.

Studies into wave crests continue at the time of writing, and may well result in future changes in the calculation of 10,000-year values. These studies include ongoing JIPs looking into crests in shallow water, and the differences between crest elevations over an area compared to a specific point.

(e) EWL: joint probability issues

The availability of long-term hindcast metocean datasets has provided valuable information on the relationships and degree of correlation between tide, surge and wave crest.

For many areas there is generally little direct correlation between extreme waves and surge heights. This is because typically the peak surge height will occur sometime after the peak wave crest. Furthermore as tides are periodic and waves are random there is no direct correlation between the two. Therefore it can be overly conservative to assume they will coincide. Nevertheless it is still theoretically possible for an extreme wave to coincide with a high tide.

There are different methods for taking account of joint probability: for example, directly extrapolating measured or hindcast water levels (analogous to extrapolating total current) or combining the components in some way. Joint criteria are discussed in more detail later in this chapter.

(f) EWL: update to 'interim method' 2005 (Leggett method)

As a result of changes to crest calculations, joint probability issues and the publication of *ISO 19901-1* in 2005, Leggett *et al.* carried our further work on 10,000-year EWL methodology. This resulted in a slight change to the previous interim method, as follows:

EWL 10,000 = Crest 10,000yr + Surge 1yr + MSL

This method still provides an acceptable result if the input data are realistic and the derivation of the component parts is done appropriately.

(g) EWL: ISO approaches for deriving EWL

With the publication of various ISO documents after the year 2000, there was an attempt to consolidate knowledge and provide information for the calculation of EWL.

ISO 19901-1 discusses the metocean aspects of EWL and, although it outlines the issues to be aware of, it does not explicitly detail a preferred approach or methodology. It highlights in particular two key issues that need to be considered (discussed earlier in this section):

- the use of a 'storms-based' approach when analysing wave data rather than a 'design sea state' approach, to ensure that short-term variability is taken account of (as discussed previously for Hmax) so that a realistic crest value is derived
- the ability to consider the joint occurrence of tide, surge and wave crest

ISO 19902 does provide a methodology for the calculation of EWL. It uses an empirical approach by taking independent design values for tide, surge and wave crest (from a metocean design report) and then combining them by reference to two equations. The choice of equation is dependent on the degree of association between the surge event and the wave crest in the extreme storm.

If the storm surge is not expected to occur at the same time as the wave crest, equation (1) applies:

$$H1 = \sqrt{(a^2+s^2+t^2)} + f \quad (1)$$

If the storm surge is expected to occur at the same time as the wave crest, equation (2) applies:

$$H2 = \sqrt{((a+s)^2+t^2)} + f \quad (2)$$

where

a is the abnormal wave crest height (10,000-year)
s is the extreme storm surge (100-year)
t is the maximum elevation of the tide relative to MSL
f is the expected sum of subsidence, settlement and sea-level rise over the design service life of the structure.

The difference in results between equations (1) and (2) can be considerable, potentially several metres in areas with high surge values. As a result operators often take a conservative approach, using equation (2), although some select a value between the two equations, particularly in areas where there is little correlation between waves and surge.

(h) EWL: implications of the changes in the derivation of EWL

Changes in the derivation of EWL and its constituent parts have had a number of consequences. These may be summarised as follows:

- With respect to the individual component parts of EWL
  - crests have increased significantly for equivalent return periods
  - however, this has been partially offset by reductions in the tide and surge components as a result of joint probability analyses.
- The changes have resulted in increases of *c.* 10% for 100-year values
- The requirement for 10,000-year values has led to a further increase in EWL compared to the previous requirement of 100-year values, despite not having to add the 1.5 m air gap. Typical differences between 100-year and 10,000-year EWL values are of the order of 25%.
- An increased EWL has potential consequences for some existing installations in that their decks may be susceptible to wave inundation.

Despite the evolution of techniques and methods, there is still no preferred option for deriving 10,000-year EWL. The *ISO 19902* equations are widely used but other approaches, including direct extrapolation of total water level data and combining crest, surge and tide (such as the Leggett method), are also used.

Finally, it needs to be appreciated that whatever method is used, derivation by extrapolation of data introduces variability in any results, depending on the software, analyses and people, as highlighted previously. It is therefore important again to ensure that appropriate expertise is utilised.

(i) EWL: terminology issues

A final and important point to mention with respect to EWL concerns clarity of terminology. There have been instances of crest elevation values being quoted in design reports when they were actually EWL values and vice versa. Clearly this can lead to the use of incorrect values in design, which could have serious consequences. Therefore, to clarify the correct terminology again:

- crest elevation should refer just to wave crest (above tide and surge)
- EWL should refer to the combination of tide, surge and crest

Design criteria for other metocean parameters

The other metocean parameters that require design values are primarily air and sea temperature and, for some areas, snow and ice information.

Air and sea temperatures

Whilst, normally, considerable air temperature data are available from sensors installed on offshore installations, there are often fewer data on sea temperature. There are, however, regional and global databases of both air and sea temperature held by various institutes and organisations, and both data and information can be obtained from them.

If there are sufficient data, a standard EVA can be carried out. However, because there is a requirement for extreme minima and maxima, any extrapolation has to take this into consideration.

Directionality and seasonality

In the previous sections on design criteria we have focused on deriving values for what is commonly called the 'all-year omni-directional' situation. This means that the values apply throughout the year and for all directions. However, there is a need to expand these values for specific requirements by deriving additional criteria for certain directions and individual months and/or seasons.

Deriving directional and seasonal criteria has to be done with care because we are dealing with a smaller proportion of the metocean database for each directional sector or month. For example, on average only one-eighth of the metocean data will be available in each direction sector for a typical 8-point compass. Analysing a smaller proportion of the database can make derivation of extreme values more challenging.

Directional criteria

The most commonly required directional criteria are for the 8 main compass points in 45° sectors centred on: North; North East; East; South East; South; South West; West; and North West.

Thus the North sector is 337.5° through to 22.5°, North East is 22.5° to 67.5° and so on. Note that there may be a requirement to centre these 8 sectors on 'platform North', i.e. the heading direction for the platform or installation. For example, if platform North is, say, 340° then the North sector would be 317.5° through to 2.5°.

There may also be requirements to further refine the directional criteria if narrower sectors are required for specific installations or activities: for example, wind turbines, optimising rig headings etc. Under such circumstances wind direction sectors may be 12 × 30° sectors or even 16 × 22.5° sectors.

An important issue with respect to directionality relates to the overall reliability of the structure requiring the criteria. As we have already provided the 50-year or 100-year value, which represents the all-year omni-directional situation, the directional criteria will need to reflect the requirement to ensure the overall probability of the extreme event is maintained.

For example, if you derive 100-year design criteria for each of the 8 compass points for an offshore structure, this means that there is a 1% probability of it experiencing the event for each direction in any one year.

This means that consideration needs to be given to selecting potentially longer return periods for certain directions to maintain the overall target probability. This can result in return periods between, say, 100 and 1,000 years for each direction.

Techniques for deriving directional criteria

Historically, two main techniques have been used routinely to derive directional criteria: the ratio method and direct extrapolation. Whilst these are still used today, a third and recent addition – the 800-year approach proposed by Forristall – is becoming more widely used.

(A) RATIO METHOD

This method uses as its starting point the 99 percentile value for each of the 8 directional sectors from an appropriate dataset. The highest value is assigned a value of 1 and the other sectors are assigned a ratio value between 0.01 and 0.99 depending on their ratio to the highest value.

Equating the sector with a ratio of 1 to the existing omni-directional design value then enables derivation of initial directional values for the other sectors by multiplying the omni-directional value with the other ratios.

Some 'uplift' in the resultant directional values is often applied by metocean specialists to smooth out any jumps between adjacent sectors and to make allowance for 'engineering judgement'. This often results in effectively increasing the return period in excess of 100 years for some directions.

Table 7.9 provides a worked example based on a 100-year omni-directional Hs of 12.0 m and a NW sector with a ratio of 1. Thus, for example, SW has a ratio of 0.84 (5.8 divided by 6.9) giving an initial directional value of 10.1 m. This has then been uplifted to 10.5 m for the final value.

(B) DIRECT EXTRAPOLATION

This is the same approach as carrying out EVA on the omni-directional data. Thus analyses are carried out for each sector in turn. However, as mentioned previously, because we are dealing with on average only one-eighth of the database for each sector, it can sometimes be difficult to derive accurate or meaningful values. This means that care is needed in selecting the criteria.

As with the ratio method this tends to result in a degree of judgement and uplift, which invariably means that each sector's 100-year value is more likely equivalent to a much longer return period, say 200–500 years.

(C) THE FORRISTALL 800-YEAR METHOD: COMPOSITE DIRECTIONAL CRITERIA

Reference has been made above of the need to ensure overall target probability of a 100-year event, and how results from both the ratio method and direct extrapolation are invariably increased. This has been taken a step further by Forristall, who has

Table 7.9 The ratio approach to directionality.

	N	NE	E	SE	S	SW	W	NW
99 percentile Hs(m)	6.2	4.1	4.0	5.2	5.1	5.8	6.4	6.9
Ratio to highest (NW)	0.90	0.59	0.58	0.88	0.74	0.84	0.93	1.0
Omni-directional Hs100								12.0
Initial directional Hs100	10.8	7.1	7.0	10.6	8.9	10.1	11.2	12.0
Final directional Hs100	11.5	9.5	9.0	11.0	10.0	10.5	11.5	12.0

suggested the use of specific 800-year criteria for each of the 8 directional sectors. This ensures the overall target probability of a 100-year event remains the same directionally and omni-directionally.

Monthly/seasonal criteria

The derivation of criteria for individual months tends to follow a similar approach to that for directional criteria. Thus the ratio method can be used for each month throughout the year as well as direct extrapolation of each month.

As well as looking at each month in isolation, it can also be useful to carry out the analyses on a rolling 3-month basis. This means deriving January criteria by analysing December–February inclusive, and February criteria by analysing January–March inclusive etc. This approach helps to smooth out the results.

Seasonal criteria are applicable to a combination of months that generally relate to more benign periods, for example:

- May–August (referred to as summer in the northern hemisphere)
- April–September (referred to as fair weather in the northern hemisphere)

In the example above, May–August criteria would be derived by analysing the four months combined.

It should be noted, however, that in some areas summer is not necessarily the most benign period. For example, worst conditions can occur during this time in West Africa, and currents can be at their strongest in the Atlantic Margin.

Joint criteria

Although the terms 'joint criteria' or 'joint probability' have been mentioned previously, primarily with respect to current profiles and EWL, this section discusses the subject in greater detail paying particular attention to winds, waves and currents.

We have already discussed the derivation of the primary design criteria, specifically Hs, mean wind speed, and current speed. The traditional approach for these criteria is to derive each parameter in isolation, i.e.

- 50-year or 100-year Hs,
- 50-year or 100-year mean wind speed
- 50-year or 100-year current speed

This approach therefore generates what are termed independent criteria. It assumes that the 100-year condition will occur at the same time for all parameters. However, this assumption is extremely unlikely because of the way that winds, waves and currents respond to extreme storm events. It therefore means that such so-called independent criteria can be overly conservative.

If we think about the occurrence of peak events for winds, waves and currents during storms, the typical sequence would be:

- the peak wind occurs first
- closely followed by the peak wave height but
- with a longer delay before the peak current speed occurs

If we try to approximate this temporally, it means that if peak wind speed occurs at time T= 0 hours, then

- the peak wave height will occur at, say, T+ 1–3 hours, and
- the peak current speed will occur at, say, T+ 6–9 hours

This leads to the concept of joint criteria or the joint probability of occurrence, i.e. trying to identify a 50-year or 100-year event rather than the individual criteria for winds, waves and currents. As waves tend to dominate the contribution to loads, in most cases it is usual to consider winds and currents that are 'associated' with the 50-year or 100-year wave. In simple terms this means trying to assess what the associated values are and sometimes trying to equate them to a (reduced) return period.

Thus, the use of joint criteria approaches can result in reductions to the associated parameters compared to independent criteria, providing a more realistic design basis and – more significantly – offering potential for major cost savings.

It should be noted that most metocean contractors now have their own software and analytical techniques for deriving joint criteria.

Approaches for deriving joint criteria

There are three main approaches that can be used in deriving joint criteria, ranging from a basic approach using simple assumptions about the combinations of parameters through to sophisticated analytical techniques. The three main approaches are:

- design rules
- design contours
- response-based design

(A) DESIGN RULES

This is a simple basic approach that can be used in the absence of anything else. The most commonly used are 'DNV rules', which allow various combinations of winds, waves and currents: for example, 100-year wave height + 100-year wind speed + 10-year current.

Table 7.10 shows various combinations of return periods for metocean data types permitted in the design-rules approach for the ultimate limit state (ULS) and for the accidental limit state (ALS).

Table 7.10 'DNV Rules': Various combinations of return periods for metocean data types.

Limit state	Winds	Waves	Currents	Water level
Ultimate limit state (ULS)	100-year	100-year	10-year	100-year
	10-year	10-year	100-year	100-year
Accidental limit state (ALS)	Return period not less than 1 year			

The limit state is a state beyond which a structure no longer satisfies the design requirements. The following categories of limit states are of relevance for structures:

- ULS = ultimate limit states
- FLS = fatigue limit states
- ALS = accidental limit states
- SLS = serviceability limit states

For example, ULS corresponds to the maximum load carrying resistance.

This approach thus allows some reduction in associated return periods – for example, 10-year currents – but does not enable the significant reductions that are possible with more sophisticated techniques. It is, therefore, a conservative approach for developing joint criteria.

(B) DESIGN CONTOURS

The traditional method to evaluate the extreme response of a structure has been to calculate the response for the 50-year or 100-year design value of a governing metocean parameter (such as wave height). Whilst this methodology is appropriate when the response is governed by a single parameter such as wave height, it is not so when responses are governed by multiple parameters.

This recognition led to the development of the iFORM contour approach in Norway (Winterstein 1993). The iFORM technique is typically used to generate contours of wave height (Hs) and associated period (Tp), where each combination of the two on a contour line has the same joint probability of exceedance. The contours are usually produced as overlays for wave scatter diagrams. The contours can then be used to assess the response of a structure or vessel to combinations of wave height and period that generate the 50-year or 100-year condition.

The iFORM method works by fitting the distribution of Tp for each Hs interval. The conditional distribution of Tp is then used to calculate pairs of Hs and Tp which have a known joint probability of exceedence.

For example, with respect to a 100-year return period, one pair is the 100-year Hs with the mean Tp for that Hs interval. Similarly, another pair is the mean Hs with the 100-year Tp. Figure 7.8 gives an example of an iFORM contour plot.

(C) RESPONSE-BASED DESIGN CRITERIA

The most recent and more sophisticated approach to the derivation of joint criteria is the response-based approach, which was originally developed by Shell in the late 1990s when they produced their load statistics module (LSM), initially for fixed steel structures.

This approach brought together both structural and metocean engineers and, with the new software, quantified the combination of metocean parameters generating a 100-year response for base shear (bs) and overturning moment (otm).

However, by virtue of its name a response-based approach means that the software needs to vary depending on the type of structure. Put simply, a fixed steel jacket or wind turbine will respond differently to a floating installation such as an FPSO. Thus, response-based criteria are installation specific.

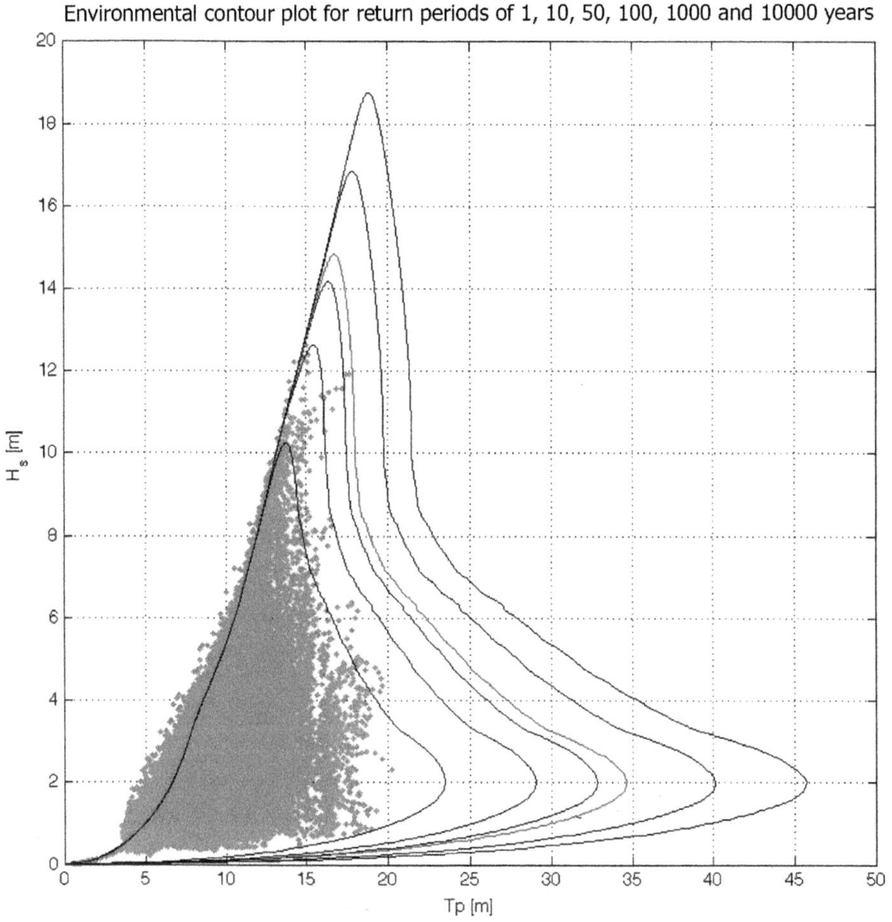

Figure 7.8 An iFORM contour plot (the red line is for the 100-year condition).

Fixed steel structures

For fixed steel structures waves tend to dominate loads: therefore the worst response tends to be the 50-year or 100-year wave plus the associated wind and current speeds.

From extensive past analyses carried out by the author in the North Sea, associated values of the wind and current for the 100-year wave condition tend to be approximately equivalent to:

- a 10-year independent wind, and
- a 1-year (or less) independent current

The significant reduction in the associated current can result in major cost savings, particularly in more shallow water depths where currents become more significant with respect to loads. However, it should be noted that these approximate North Sea associated values will not necessarily apply in other parts of the world.

Floating structures

For floating structures the situation is different and far more complicated, in particular due to motions such as heave, pitch and roll. This means additional metocean parameters are involved in any analyses: for example, wave period and the directionality of winds, waves and currents.

As a result it is interesting to note that the worst response in terms of load can be attributed to different return period combinations of parameters and not necessarily just to 100-year metocean events. For example, a 100-year response may be a result of say:

- a 1-year Hs from the SW
- a 50-year Tp
- a 10-year wind from the NE and
- a 50-year current from the NW

Figure 7.9 illustrates the impact of directional differences between winds and waves on vessel roll. It can be seen that maximum roll values occurred when winds and waves were offset by circa 135°.

Additional analytical presentations

In the operational statistics section of this chapter, reference was made to a range of JFDs that may be required. Because many of these are straightforward analyses

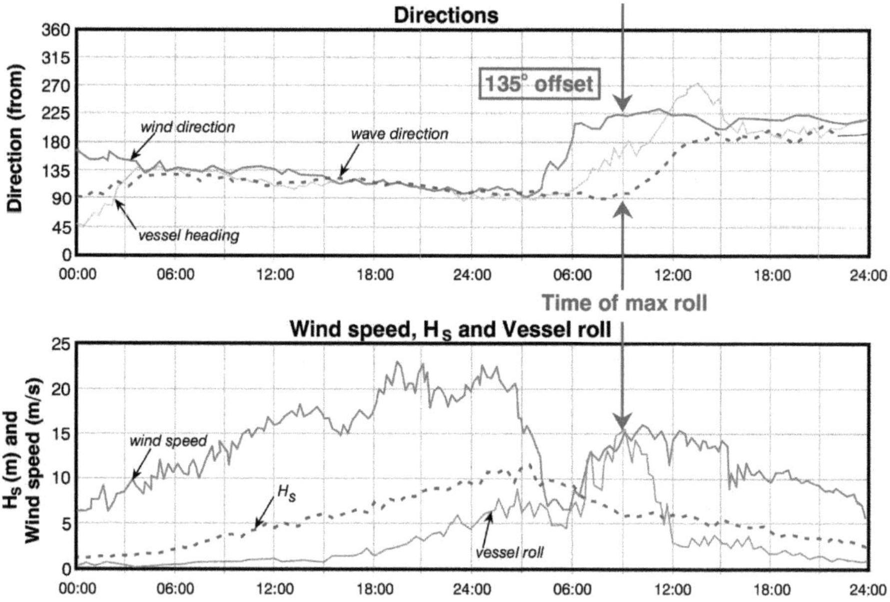

Figure 7.9 Time series of winds, waves and vessel roll.

of past data they can be perceived as operational statistics. However, with hindcast databases of 40+ years these outputs and presentations are alsdo invaluable for design requirements.

There are also a few JFDs that are produced with input from design criteria information. The most commonly required example is the individual wave height distribution. This can be presented in a number of different ways to assist with fatigue analysis but all presentations tend to be associated with showing information on the number of waves by directions with associated periods relating to 30 years, which in many cases relates to the lifetime of a structure.

Chapter 8
Metocean reports: Contents, format, presentations

The previous chapter focused on the analysis of metocean data to derive operational statistics and design criteria. It also touched on a few of the ways in which the results of those analyses are presented. This chapter provides more detail on the contents, format and types of presentation required to produce metocean reports containing operational statistics and design criteria.

Although in many cases there will be a requirement for separate comprehensive operational statistics or design criteria reports, for some applications there will be a requirement to produce one overall report containing both operational statistics and design criteria, often as separate volumes. Furthermore, for some applications there may be a requirement for a report that focuses on specific issues: for example, tow-route criteria or response-based criteria.

Whatever information is presented, the contents should contain a mixture of textual or descriptive information together with presentation material usually in the form of tables, figures or graphs.

Descriptive information

Although end-users of metocean reports are usually only interested in numerical values and presentations, it is important to ensure that appropriate descriptive text is included to show how values have been derived, what data have been used, together with a location map, glossary and any other pertinent information – for example, the standards adhered to.

It is also important to ensure that presentations such as figures and tables contain appropriate text. This includes such things as location, water depth, data sources including start/end dates and number of data points, and any other salient information.

It can also be useful to include text examples of how to interpret certain presentations.

Operational statistics: presentational material

There is a wide range of potential tables and figures for presenting operational statistics. This – together with the fact that for some presentations there can be analytical output for different probability levels – means there is the potential for generating a considerable volume of material. It is therefore beneficial for both providers and end-users to discuss specific requirements and uses before finalising the exact presentations needed.

A listing of typical operational statistics presentations is:

- Tables of 95%, 50%, 25%, 10%, 1% and 0.1% monthly percentage exceedance values for Hs, 10-min mean wind speed, surface and near seabed current speed and near seabed current.
- Tables of monthly percentage exceedance values for 10-min mean wind speeds (in 2.5 m/s intervals) and Hs values (0.5 m intervals). Table 7.6 gives an example of a monthly percentage exceedence table for Hs.
- Wind rose (10-min mean): all-year, seasonal and monthly.
- Wave rose (Hs): all-year, seasonal and monthly.
- Persistence/duration statistics of 10-min mean wind speed
  - thresholds every 2.5 m/s from 2.5 m/s to 25 m/s for less than these values 2.5 to 25.
  - thresholds every 2.5 m/s from 2.5 m/s to 25 m/s for greater than these values 2.5 to 25.
- Persistence/duration statistics for Hs
  - thresholds every 0.5 m from 1 m to 5 m for less than these values 1 to 5.
  - thresholds every 0.5 m from 1 m to 5 m for greater than these values 1 to 5.
- Monthly air temperatures for various exceedance values (95%, 50%, 25%, 10%, 1% and 0.1%) plus minimum and maximum recorded values.
- Monthly sea surface temperatures for various exceedance values (95%, 50%, 25%, 10%, 1% and 0.1%) plus minimum and maximum recorded values.
- Monthly sea temperature profiles for surface, mid-depth and seabed (minimum, mean and maximum values).
- Other metocean parameters, including visibility, rainfall, snow, cloud cover and humidity, normally presented as monthly frequency distributions.

Note that exceedance and persistence tables are typically presented as average values from the *n*-year database. Other probability levels may need to be included such as P10 and P90. An example of a persistence table is provided in Chapter 7 (see Table 7.8).

Note also that more sophisticated 'persistence' output can be derived from detailed weather windows analysis: for example, to quantify start/end times, multi-threshold operations etc. This is also described in Chapter 7.

Design criteria reports: presentational material

As with operational statistics there is a wide range of potential presentations for design criteria. Because of this it is often useful to provide an overview of the key criteria in a summary table at the beginning of a report.

The example provided in Figure 8.1 includes 1-, 10-, 50- and 100-year criteria for the all-year, April–September and May–August periods for a North Sea location.

The remaining figures in this chapter are described by data type. Unless otherwise stated, with respect to the terminology:

- 'return period(s)' means 1, 10, 50 and 100 years
- 'all-year' means January–December
- 'summer' means May–August (in the northern hemisphere)

All YEAR

	RETURN PERIOD (YEARS) 1	10	50	100
WIND SPEED 10 m asl [m/s]				
V (1-hr)	24.9	29.0	31.8	33.0
V (10-min)	27.2	31.9	35.1	36.5
V (1-min)	30.1	35.6	39.4	41.0
Gust (3-sec)	33.9	40.4	45.0	46.9
WAVE HEIGHT [m] & PERIOD [s]				
Hs	6.4	8.0	9.0	9.5
Tz (central)	8.1	9.1	9.7	9.9
Tp (central)	11.6	13.0	13.8	14.2
Hmax(3-hr)	10.9	13.2	14.7	15.3
Tass (central)	10.4	11.6	12.4	12.7
CURRENT PROFILE [m/s]				
1.00d	0.86	0.95	1.00	1.03
0.75d	0.86	0.95	1.00	1.03
0.50d	0.86	0.95	1.00	1.03
0.30d	0.80	0.88	0.93	0.95
0.10d	0.68	0.75	0.80	0.81
0.05d	0.62	0.68	0.72	0.74
1m asb	0.59	0.65	0.68	0.70

APRIL to SEPTEMBER

	RETURN PERIOD (YEARS) 1	10	50	100
WIND SPEED 10 m asl [m/s]				
V (1-hr)	18.3	21.4	23.4	24.3
V (10-min)	19.8	23.2	25.5	26.5
V (1-min)	21.7	25.5	28.2	29.3
Gust (3-sec)	24.1	28.5	31.6	33.0
WAVE HEIGHT [m] & PERIOD [s]				
Hs	4.8	6.0	6.8	7.2
Tz (central)	7.1	7.9	8.4	8.7
Tp (central)	10.1	11.3	12.1	12.4
Hmax(3-hr)	8.6	10.4	11.6	12.1
Tass (central)	9.1	10.1	10.8	11.1
CURRENT PROFILE [m/s]				
1.00d	0.78	0.86	0.91	0.93
0.75d	0.78	0.86	0.91	0.93
0.50d	0.78	0.86	0.91	0.93
0.30d	0.72	0.79	0.84	0.86
0.10d	0.62	0.68	0.72	0.73
0.05d	0.56	0.61	0.65	0.66
1m asb	0.53	0.58	0.62	0.63

All YEAR

	RETURN PERIOD (YEARS) 100 Minimum	100 Maximum
TEMPERATURE [°C]		
Air Temp.	-6.5	24.5
Sea Surface Temp.	0.0	20.5
Sea Bed Temp.	1.5	16.5

MAY to AUGUST

	RETURN PERIOD (YEARS) 1	10	50	100
WIND SPEED 10 m asl [m/s]				
V (1-hr)	16.2	18.9	20.7	21.5
V (10-min)	17.4	20.4	22.5	23.3
V (1-min)	19.0	22.3	24.7	25.7
Gust (3-sec)	21.0	24.9	27.6	28.7
WAVE HEIGHT [m] & PERIOD [s]				
Hs	3.6	4.5	5.1	5.3
Tz (central)	6.1	6.8	7.3	7.5
Tp (central)	8.7	9.7	10.4	10.7
Hmax(3-hr)	6.5	8.0	8.9	9.3
Tass (central)	7.8	8.7	9.3	9.5
CURRENT PROFILE [m/s]				
1.00d	0.66	0.73	0.77	0.79
0.75d	0.66	0.73	0.77	0.79
0.50d	0.66	0.73	0.77	0.79
0.30d	0.61	0.68	0.72	0.73
0.10d	0.53	0.58	0.61	0.63
0.05d	0.48	0.52	0.55	0.57
1m asb	0.45	0.50	0.53	0.54

All YEAR

	RETURN PERIOD (YEARS) 100	10000
EXTREME WATER LEVEL [m]		
Extreme Water Level (Relative to LAT)	14.6	18.5

Figure 8.1 A summary criteria table.

- by 'wave heights' is meant Hs and Hmax (but note that crest can also be considered a wave height)
- 'wave periods' means a range (lower, mid and upper) of Tz, Tp and Tass

Wind presentations

- Tables of omni-directional all-year and seasonal speeds for various durations from 24 hours to 3 seconds for all return periods
- Tables of directional all-year and seasonal speeds for various durations from 24 hours to 3 seconds for all return periods
- Table of vertical wind profile 10–150 m ASL in 10 m intervals

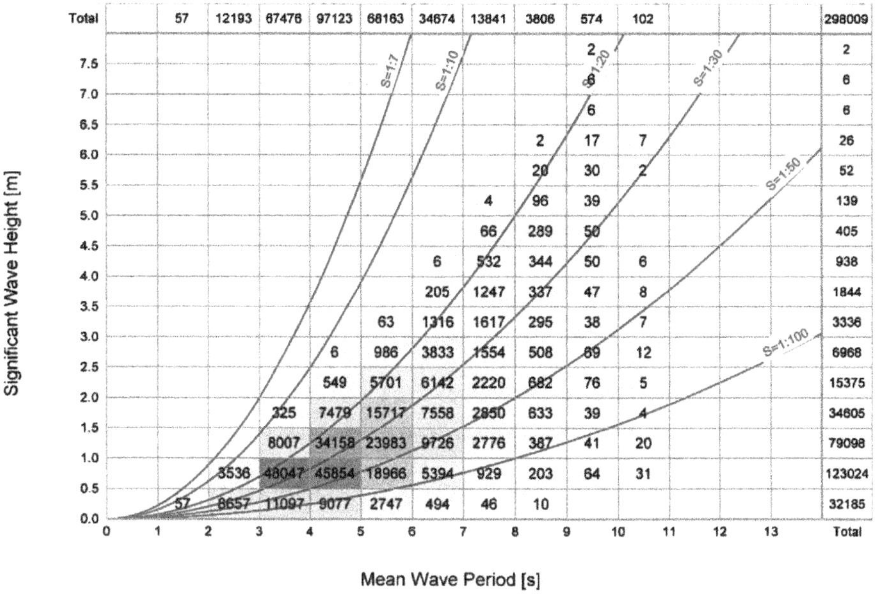

Figure 8.2 A wave scatter diagram (Hs vs Tz).

Wave presentations

- Joint frequency tables of omni-directional all-year and seasonal wave heights and wave periods (Hs vs Tz, Hs vs Tp) for all return periods, both as numbers of observations and percentages. These presentations are normally referred to as wave scatter diagrams (see Figure 8.2)
- Joint frequency tables of directional all-year and seasonal wave heights and wave periods (Hs vs Tz, Hs vs Tp) for all return periods, both as numbers of observations and percentages
- Joint frequency tables of the total number of individual waves by direction by height (H) and period (T) for 1 year and 30 years in 1 m intervals for H and in 1 second intervals for T. These presentations give an indication of the total number of waves that will pass through a structure in a typical year and during its lifetime, and are sometimes referred to as fatigue tables

Current presentations

- Tables of directional all-year and seasonal total current profiles for all return periods. The levels in the water column are normally expressed relative to the depth (d): for example, 1.0d (surface), 0.5d (mid-depth), and 0.01d (seabed)

Water level presentations

- Table of tidal levels from LAT to HAT
- Table of positive and negative surge levels for all return periods plus 10,000-year value

- Table of positive and negative SWL for all return periods plus 10,000-year value
- Table of 100-year and 10,000-year EWL with values for constituent parts as appropriate, i.e. wave crest, tide and surge

Temperature presentations

- Table of 100-year extreme minimum and maximum air temperatures at 10 m ASL
- Table of 100-year sea temperature profiles (surface, mid-depth and seabed)

Other presentations (if relevant)

- Tables of ice and snow accumulations
- Table of structure-specific all-year omni-directional joint criteria for winds, waves and currents

Chapter 9
Climate variability and data trends

The Earth's climate has always been subject to variability or change due to a number of factors which, broadly, fall into two main areas:

- internal forcing mechanisms, such as ocean–atmosphere variability
- external forcing mechanisms, such as solar radiation, the Earth's orbit, volcanoes and plate tectonics.

The variability resulting from these mechanisms may be referred to as natural variability.

There is also variability or change in the climate due to the influence of human activity, referred to as anthropogenic change. Many of these activities – for example, the increase in CO_2 emissions – are tending to have a warming effect on the planet.

We therefore have a combination of natural variability and anthropogenic change, which sometimes act together to 'enhance' any changes but which can occasionally also appear to cancel each other out.

The increase in global air temperatures over the last 170 years or so is illustrated in Figure 9.1. It shows information from six sources presented as anomalies referenced

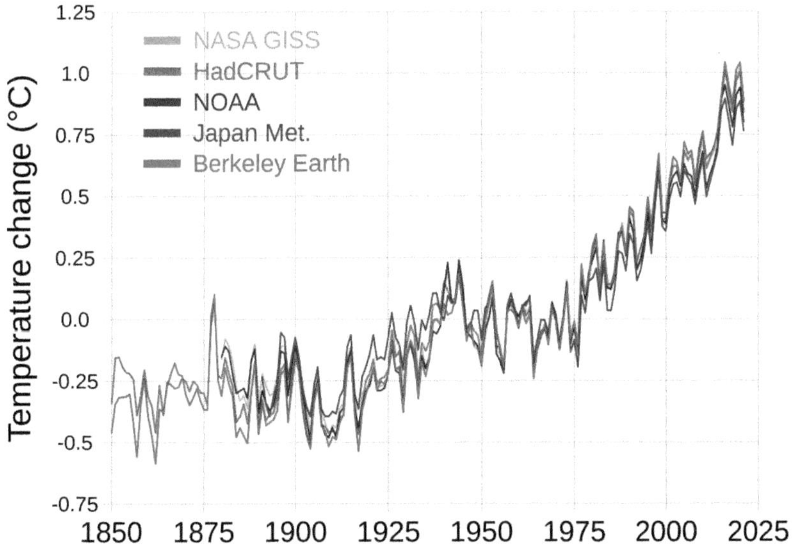

Figure 9.1 Global average temperature change (courtesy RCraig09, CC BY-SA 4.0 <https://creativecommons.org/licenses/by-sa/4.0>, via Wikimedia Commons).

to the years 1951–1980. Note in particular the relatively steep increase since 1980, and the consistency of the results between the six sources.

Climate change is a huge subject in its own right and, as such, this chapter will not discuss the wide range of issues associated with it, in particular external forcing mechanisms and anthropogenic factors. These are addressed by numerous other documents, particularly those issued by the Inter-Governmental Panel on Climate Change (IPCC). What this chapter will do is:

- provide some information on internal forcing mechanisms, in particular ocean–atmosphere variability
- illustrate how such natural variability can impact metocean conditions such as the wave climate

Clearly any potential change or long-term trend – in for example wave conditions – can have implications for the derivation of operational statistics and design criteria. This is because it may not necessarily be the case that deriving 50-year or 100-year values from data for the past 40 years will be appropriate for the next 40 years.

Ocean–atmosphere variability

The ocean and atmosphere can work together to spontaneously generate internal climate variability that can persist for years or decades at a time. Examples of this type of variability include:

- the El Niño Southern Oscillation (ENSO)
- the Pacific Decadal Oscillation (PDO)
- the Atlantic Multi-Decadal Oscillation (AMO)
- the North Atlantic Oscillation (NAO)

Further information is provided on two of these: ENSO and NAO.

El Niño Southern Oscillation (ENSO)

Perhaps the most well-known example of ocean–atmosphere variability is El Niño and its 'opposite', La Niña, which occur in the eastern Pacific.

El Niño is characterised by major oceanic warming that spreads westwards. It is such a dominant effect that it has major climate implications around the world. La Niña has the opposite effect, characterised by cooling. It is not such a dominant feature as El Niño but can still have an impact around the world.

Figure 9.2 shows how the sea temperature in the Pacific changes as a result of El Niño and La Niña. In particular, note the warming of the eastern Pacific characterised by the 'red plume' spreading westwards during an El Niño.

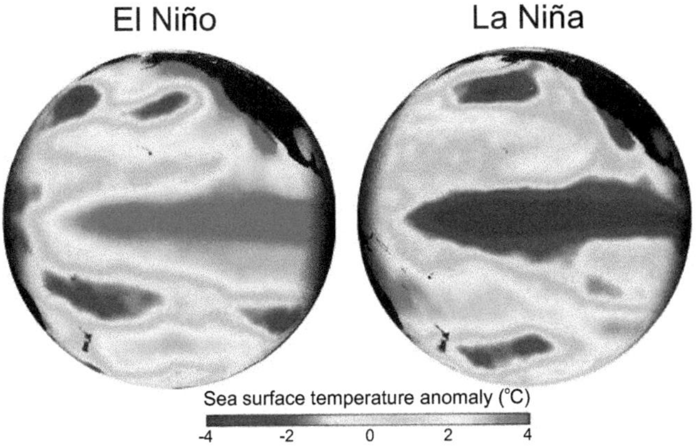

Figure 9.2 Sea temperature anomalies associated with El Niño (left) and La Niña (right) (courtesy of NOAA).

As mentioned, El Niño causes changes to the weather in other parts of the world. Figure 9.3 gives a general overview of some of these changes for the period December–February ... although all these impacts are not necessarily generated by each warm- or cool-water event.

It should therefore be appreciated that because El Nino/La Nina cause different changes on a regional basis the impact of climate variability is not always uniformly global. This also applies to the anthropogenic contribution to climate change in that although a global rise in temperature has been observed there is variability on a regional basis.

North Atlantic Oscillation (NAO)

The NAO is an index based on pressure differences between Iceland and the Azores. When the index is positive or high, the weather in NW Europe is characterised by more storms, westerly winds and mild/wet conditions; conversely when the index is negative or low, the opposite tends to occur with fewer storms, more easterly winds and colder conditions. Figure 9.4 illustrates these two different situations.

The NAO Index has been calculated since the middle of the 19th century and Figure 9.5 shows how it has varied since that time. The red columns indicate periods when the index was high and conversely the blue show periods when it was low.

It can be seen that periods of high or positive NAO tend to dominate, and that over the last 40 years less than 10 of them have been negative.

It would obviously be extremely useful to be able to predict the future trends of indices such as the NAO, and even more so to relate them to expected metocean conditions. Although this is not currently possible with a high degree of accuracy, there have been investigations into the relationship between wave conditions and the variability of the NAO. This is referred to in the next section.

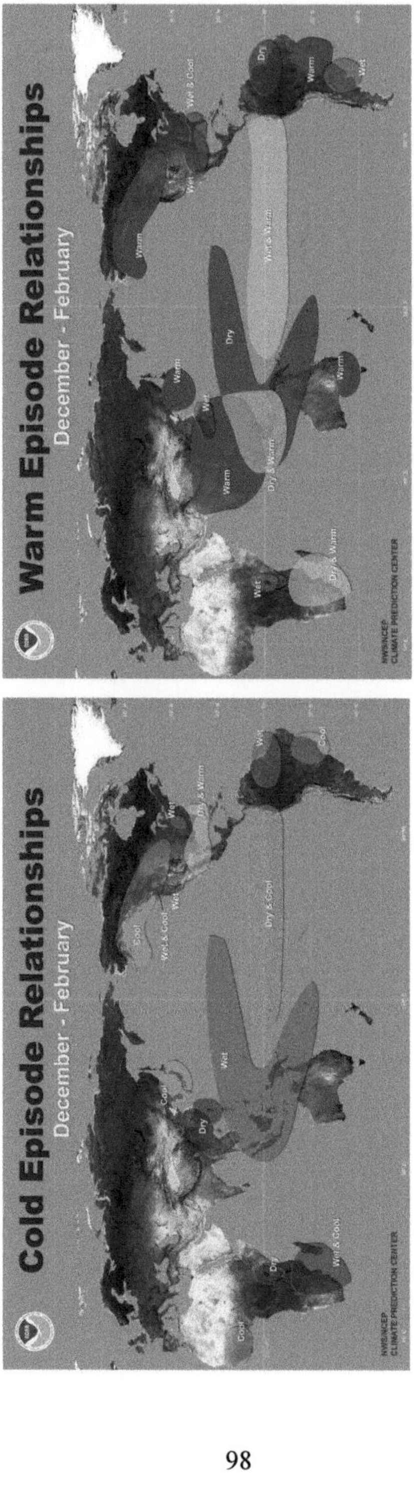

Figure 9.3 El Niño global effects December–February (courtesy of NOAA/NCEP/CPC).

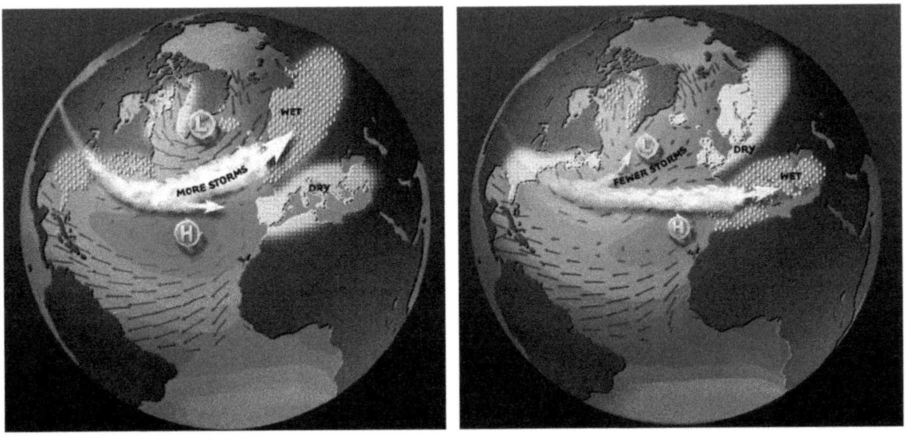

Figure 9.4 High NAO (left) – more westerlies, more storms, milder and wetter; low NAO (right) – more easterlies, fewer storms, colder and drier.

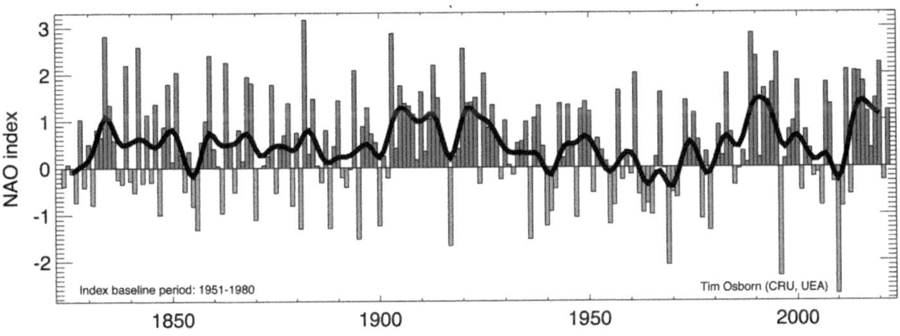

Figure 9.5 NAO index (courtesy of Professor T. Osborn, Climatic Research Unit, School of Environmental Sciences, UEA).

Wave climate trends

The variability of indices such as the NAO has the potential to impact metocean conditions over time. For example, it has been shown by the author that trends in wave heights in the North Sea can be linked with the NAO.

This can be illustrated by reference to wave data from the northern North Sea, which are available for the period 1964–2010. Figure 9.6 shows a plot of mean Hs during the winter months October–March. The red columns refer to the period October–December whereas the blue columns refer to the period January–March. The plot is overlaid with 5-year running means for each of these periods.

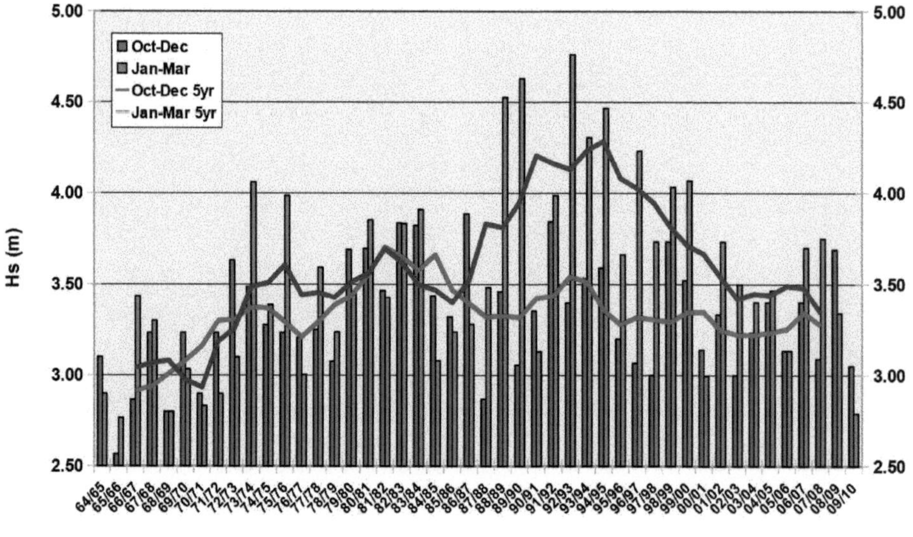

Figure 9.6 Mean Hs, winter months northern North Sea 1964–2010

There are several things to note from Figure 9.6:

- the general similarity in conditions between the period October–December and the period January–March from 1964 to the mid-1980s accompanied by a general increase in the severity of mean Hs from *c.* 3 m to 3.5 m
- a major change in the mid-/late 1980s, denoted in particular by:
  - a significant increase in mean Hs in the period January–March (up to *c.* 4.5 m)
  - whereas conditions start to decrease slightly in the period October–December
- a general reduction in mean Hs since the late 1990s; however, values have been increasing again since 2010

It should also be noted that the significantly increased storminess in the period from the mid-/late 1980s to the mid-1990s coincides with a strong positive anomaly for the NAO. This anomaly is evident on the NAO Index plot (Figure 9.5).

Notwithstanding any relationship with the NAO, trends in the wave data pose challenges when analysing data to derive operational statistics and design criteria.

In this respect it is important to consider the time period to be used to carry out any analyses, particularly EVA, to derive extreme values. For example:

- using data from the 1960s and the 1970s results in 100-year extreme values in the northern North Sea of 14–15 m, whereas
- using data from the most severe period (1988–1995) results in 100-year extreme values of 17–18 m

Chapter 10
Metocean conditions around the world

Although all the world's seas and oceans experience the effects of winds, waves, currents and other metocean phenomena, there can be considerable variability in the strength and occurrence of specific parameters and features on a regional basis.

Providing detailed descriptions of metocean conditions in different parts of the world is beyond the scope of this book. Instead this section provides a brief overview of some of the key processes and features, and compares some examples of operational and extreme values.

Processes and features in different parts of the world

Table 10.1 lists some of the dominant processes and features in several of the key areas where O&G operations take place. It should be noted that some of these processes and features have already been discussed previously and others are covered in Appendix 2.

The areas may be classified into one of three different climatic regions – temperate, tropical and cold – as follows:

- temperate: North Sea and Atlantic Margin
- tropical: Gulf of Mexico, West Africa, South China Sea
- cold: Arctic, North Caspian

The different processes and features outlined in Table 10.1 can result in significant variability in operational statistics and design criteria globally. Some examples of this are provided in Tables 10.2 and 10.3.

Table 10.1 Dominant processes and features of some key areas for O&G operations.

Area	Metocean processes and features
North Sea	Extra-tropical depressions, tidal currents
Atlantic Margin	Oceanic currents, eddies, contraflows between surface and seabed, sea temperatures at depth<0°C
Gulf of Mexico	Hurricanes, Loop Current, eddies
West Africa	Long period swell, line squalls, river outflows
South China Sea	Typhoons, monsoons, internal waves
North Caspian	Winter ice, summer heat, changing sea level
Arctic	Different types of ice, summer thaw, daylight issues

Table 10.2 Typical mean and maximum Hs statistics for three key areas of the world.

	Mean Hs (m)		Maximum Hs (m)	
	January	July	January	July
Atlantic Margin	4.5	2	14	7
West Africa	1	2	2	3.5
South China Sea	2	1	6	6

Table 10.3 Indicative 100-year extreme values for nine locations around the world.

Region	10-min mean wind speed	Hs	Hmax	Tp	Surface current	Minimum seabed temperature
	m/s	m	m	S	m/s	°C
Temperate						
Southern North Sea	35	9	16	14	1.5	4
Northern North Sea	45	16	30	17	1	3
Atlantic Margin	45	18	34	18	2	−2
Tropical						
Gulf of Mexico	45	15	27	15	2.5	3
Nigeria	30	4	7	18	1.5	4
South China Sea	30	7	13	13	2	4
Cold						
North Caspian	30	4	7	8	1	−2
Barents Sea	40	15	27	17	1	−1
Sakhalin	30	10	18	14	2	−2

With respect to offshore renewables (OR) most of the activities and developments relate to shallow nearshore and coastal areas in regions with suitable wind and wave climates. Specific metocean phenomena and issues in these zones are discussed in more detail in Appendix 2. It should also be noted that, with the majority of OR developments taking place in temperate latitudes, the key metocean processes and features tend to be dominated by extra-tropical depressions, strong currents and large tidal ranges.

Global variability of Hs operational statistics

Table 10.2 presents some typical values for mean and maximum Hs values from hindcast model output for the months of January and July for three of the areas mentioned: the Atlantic Margin, West Africa and the South China Sea.

There are a few key points to note from the Table 10.2:

- the severity of conditions in the Atlantic Margin compared to the other areas
- the expectation of worse conditions in the 'winter months' does not always apply: for example, West Africa
- the similarity in the mean Hs value for 'summer' between the Atlantic Margin and West Africa

Global variability of extreme values

Table 10.3 summarises indicative 100-year extreme values for each of the three climatic zones, temperate, tropical and cold. For each zone, three locations/areas have been selected.

Clearly these values are not applicable universally throughout each area and, in this respect, site-specific criteria will always be required for any potential new development.

The key points to note from Table 10.3 are:

- the severity of conditions in the Atlantic Margin, making it one of the most hostile places globally for exploration and production activity
- the variability of conditions in different tropical regions: for example, contrast the Gulf of Mexico with Nigeria
- the variability of conditions in the cold regions: for example, contrast the Barents Sea with North Caspian
- the negative values for minimum seabed temperature in some regions

More detail on regional metocean conditions can be found in *ISO 19901-1* annexes.

References

Forristall, G.Z. (1978) On the statistical distribution of wave heights in a storm. *Journal of Geophysical Research* 83 (C5), pp. 2353–2358.

Forristall, G.Z. (2000) Wave crest distributions: observations and second order theory. *J. Phys. Oceanogr.* 30 (8), pp. 1931–1943.

Forristall, G. (2007) *Methods for Calculating Maximum Wave and Crest Heights: A Review.* Forristall Ocean Engineering, Inc.

Forristall, G.Z. (2008) How should we combine long and short term wave height distributions? Proc 27th Int. Conf. Offshore Mech. Arctic Eng (2008) (OMAE2008–58012). Estoril, Portugal, 15–20 June 2008.

Forristall, G.Z. (2008) On the use of directional wave criteria. *Journal of Waterway, Port, Coastal & Ocean Engineering*, 130 (5), pp. 272–275.

Tromans, P.S. and Vanderschuren, L. (1995) Response-based design conditions in the North Sea: Application of a new method. OTC 95, 27th Annual Offshore Technology Conference, 1–4 May 1995; Houston, Texas, USA. Proceedings. Vol 1, ppr OTC 7683.

Winterstein (1993) Environmental parameters for extreme response: inverse FORM with omission factors – Steven Winterstein Jan 1993 Probability-Based Engineering.

Appendix 1
Metocean standards, policies and guidelines

This appendix provides information on relevant standards, policies and guidelines. It is structured in three main sections:

- ISO/IEC standards
- data collection and systems
- certification agencies, governmental departments, and others

1. ISO/IEC standards

ISO standards

There are 4 ISO documents applicable to metocean. Of these, *ISO 19901-1* is specifically a metocean standard, whereas the other three – *ISO 19902, 19905-1* and *19906* – only refer to metocean. These standards were originally produced for the O&G industry; however, the metocean standard *ISO 19901-1* is also applicable to OR.

ISO 19901-1:2015 Petroleum and Natural Gas Industries – Specific Requirements for Offshore Structures Part 1: Metocean Design and Operating Considerations

ISO 19901-1:2015 gives general requirements for the determination and use of metocean conditions for the design, construction and operation of offshore structures of all types used in the petroleum and natural gas industries. In this respect it is the definitive metocean standard for the O&G sector. As noted above, however, it is equally applicable to OR.

The requirements are divided into two broad types:

- the determination of metocean conditions in general, together with the parameters that are required to adequately describe them
- the characterisation and use of metocean parameters for the design, construction or operation of offshore structures

The conditions and parameters discussed are:

- extreme and abnormal values of metocean parameters that recur with given return periods that are considerably longer than the design service life of the structure
- long-term distributions of metocean parameters, in the form of cumulative, conditional, marginal or joint statistics
- 'normal' (typical) conditions that are expected to occur frequently during the design service life of the structure

Metocean parameters are applicable to:

- the determination of actions for the design of new structures and the assessment of existing ones
- the site-specific assessment of mobile offshore units
- the determination of limiting conditions, weather windows, actions and action effects for fabrication, transportation, installation and decommissioning
- the operation of the platform, where appropriate

ISO 19902:2007 Petroleum and Natural Gas Industries - Fixed Steel Offshore Structures

ISO 19902:2007 specifies requirements and provides recommendations applicable to all fixed steel offshore structures for the petroleum and natural gas industries. It contains requirements for the planning and engineering of the following tasks:

- design, fabrication, transportation and installation of new structures as well as their future removal
- in-service inspection and integrity management of both new and existing structures
- assessment of existing structures
- evaluation of structures for re-use at different locations

Although this standard is specifically for structural engineers, it does contain a method for the derivation of extreme water level (EWL). (See Chapter 7 for further information.)

ISO 19905-1:2016 Petroleum and Natural Gas Industries – Site-Specific Assessments of Mobile Offshore Units – Part 1: Jack-Ups

ISO 19905-1:2016 specifies requirements and guidance for the site-specific assessment of independent leg jack-up units for use in the petroleum and natural gas industries.

There are some specific metocean requirements for jack-up assessments in this document.

ISO 19906:2010 Petroleum and Natural Gas Industries – Arctic Offshore Structures

ISO 19906:2010 specifies requirements and provides recommendations and guidance for the design, construction, transportation, installation and removal of offshore structures, related to the activities of the petroleum and natural gas industries in arctic and cold regions.

Reference to arctic and cold regions in *ISO 19906* is deemed to include both the Arctic and other cold regions that are subject to similar sea ice, iceberg and icing conditions.

IEC standards

There is a plethora of IEC standards of use in the OR sector. IEC has produced a suite of standards under 61400. Three of the most important containing metocean elements are *Wind Energy Generation Systems – Part 3-1*; *Wind Energy Generation Systems – Part 3-2*; and *Technical Specification IEC TS 62600-2*

Wind Energy Generation Systems – Part 3-1

IEC 61400-3-1:2019 Wind Energy Generation Systems – Part 3-1: Design Requirements for Fixed Offshore Wind Turbines outlines design requirements for fixed offshore wind turbines, including assessment of the external conditions at the site. It specifies essential design requirements to ensure the engineering integrity of the turbines.

Wind Energy Generation Systems – Part 3-2

IEC 61400-3-2:2020 (Draft) Wind Energy Generation Systems – Part 3-2: Design Requirements for Floating Offshore Wind Turbines is a new document complementing *61400-3-1* to address the issue of floating offshore wind turbines, and is due for publication in 2024.

Technical Specification IEC TS 62600-2

Technical Specification IEC TS 62600-2:2019. Marine Energy – Wave, Tidal and Other Water Current Converters – Part 2: Design Requirements for Marine Energy Systems provides the essential design requirements to ensure the engineering integrity of wave, tidal and other water current energy converters, referred to as marine energy converters (MECs), for a specified design life. It includes requirements to ensure the structural viability of the device under site-specific external environmental conditions. This document applies to structures that are either floating or fixed to the seafloor or shore. It also applies to structures that are unmanned during operational periods.

2. Data collection and offshore systems

There are six stand-alone documents relating to data collection, offshore systems and procedures. Although five of these documents have been developed by the O&G industry, the first document – *OGP 447* – is also applicable to OR.

OGP 447 Health, Safety and Environmental Guidelines for Metocean and Arctic Surveys (2011) was produced by the IOGP Metocean Committee and details health, safety and environmental guidelines for carrying out metocean measurement campaigns. It also includes information on Arctic surveys.

The purpose of the document, *Standard Measuring Equipment for Helideck Monitoring System (HMS) and Weather Data – HCA, Bristow, Bond and CHC: Rev 8c 2012*, is to ensure uniformity of readings/registration of helideck movement and weather conditions. It is an agreement between Bristow Group (Bristow Norway and Bristow UK), Bond Offshore Helicopters, and CHC (CHC Norway and CHC UK) and applies to all moving helidecks operating on the UK and Norwegian continental shelves. Further intentions are to establish national and international standards based on the contents of this document.

CAP437 Offshore Helicopter Landing Areas 2016 (8th Edition) – CAA Publication details requirements for metocean systems and reporting in connection with helicopter operations worldwide for helicopters registered in the UK.

There are two other documents containing relevant information: *CAP1145* and *CAP746*. The former in particular has sections concerned with using forecast information rather than measured data for flight planning.

Adverse weather documents for O&G companies

Many O&G companies have internal documents that detail procedures to be adopted in the event of expected adverse weather conditions. Such documents are usually structured by main operational activity (such as drilling, helicopter flights, supply vessels etc.) and for each there will be a traffic light system (red, amber, green) for the various sub-activities linked to metocean conditions.

For example, the following traffic light system tends to apply in the North Sea for significant wave height (Hs) with respect to helicopter operations:

- Hs<5.5 m = green (operations can proceed)
- Hs 5.5–7 m = amber (operations only proceed if Hs will not exceed 7 m)
- Hs>7 m = red (no operations)

Obviously, these values will vary in different parts of the world.

NORSOK Standard N-002: Collection of Metocean Data, Edition 2 (October 2010)

This standard was developed by the Norwegian petroleum industry to present the functional requirements and common principles for the collection of metocean data on offshore installations. It was formerly a stand-alone annex to the 2005 edition of *ISO 19901-1* but has been removed from the 2015 edition. Guidance on data collection is now included in the ISO document. This NORSOK standard is therefore no longer required.

Metocean Procedures Guide for Offshore Renewables 2018 – IMarEST Offshore Renewables Special Interest Group (ORSIG)

This document gives a brief overview of metocean activities to support OR projects.

3. Certification agencies and governmental departments

There are various certification agencies and governmental departments worldwide that provide a range of documentation appropriate to metocean standards, policies and guidelines. I have selected two organisations that are widely referred to in the North Sea: DNV GL and the UK Health and Safety Executive (HSE).

DNV (Det Norske Veritas)

DNV has produced a considerable number of guidance documents and standards for the OR and O&G sectors. These include:

- *Recommended Practice DNV-RP-C205: Environmental Conditions and Environmental Loads (DNV 2017).* This document gives guidance for modelling, analysis and the prediction of environmental (metocean) conditions as well as guidance for calculating loads acting on structures, limited to those due to wind, wave and current.
- *Standard DNVGL-ST-N001: Marine Operations and Marine Warranty (DNV 2016).* This standard provides design and operational requirements for the temporary phases in the development of offshore assets, including transportation.

A number of subject-specific standards have also been produced during the period 2016–2018 for fixed and floating wind turbines. These include *DNVGL-ST-0119, DNVGL-ST-0126* and *DNVGL-ST-0437*.

UK Health and Safety Executive (HSE)

The HSE has produced a number of reference documents and reports that address various metocean issues and considerations. Four documents are listed below, in chronological order. It should be noted that none of them can be classified as a standard, although most do contain useful information.

Offshore Technology Report (OTR) 2001/010: Environmental Considerations (2001)

This document is a re-issue of part of the 1990 edition of the *UK Department of Energy (DEn) Guidance Notes*, which in its day was a recognised standard.

At this point it is worth providing some historical background. The UK Department of Energy provided guidance on design, construction and certification of offshore structures between 1974 and June 1998. This guidance was first published to support the *Offshore Installations (Construction and Survey) Regulations 1974 (SI 1974/289)* by providing a consistent basis for the certification of offshore installations. The guidance was regularly updated, and the 4th and final edition was published in 1990.

The document was generally referred to as the *Guidance Notes*, although sometimes as 'the 4th edition'. Section 11 of that document was entitled 'Environmental (*Metocean*) Considerations' and included contour maps of indicative values of parameters (such as Hs) with a 50-year return period, this return period being selected in accordance with the requirements of SI 1974/289 at that time.

HSE withdrew the 4th edition in June 1998 at the end of the certification regime, but re-published Section 11 as *Offshore Technology Report 2001/010.* However, on re-publication additional warning text was added as follows:

> It should be noted that the technical content of the 'Guidance' has not been updated as part of the re-formatting of the OTO publication… The user… must therefore assess the appropriateness and currency of the technical information for any specific application.

This means that this document should be used with caution as it does not contain updated information on contour maps, techniques and methods.

Research Report 392: Wave Mapping In UK Waters (2005)

This report contains a contour map of 100-year significant wave heights (Hs) for the North Sea and surrounding waters, which updated the map provided previously in *OTR 2001/010*.

OSD Technical Policy Relating to Extreme Weather Hazards (2005)

This document is concerned with the ability of duty holders (operators) to demonstrate that structures have a low probability of catastrophic failure when subjected to extreme environmental (metocean) actions.

It emphasises that primary control for extreme weather is achieved by the ability of structures to withstand extreme loads through adequate design with suitable safety factors, i.e. the 100-year return period, and a subsequent check to a longer return period, i.e. 10,000 years.

The preferred 'inherently safe' situation for fixed and jack-up installations is one where an air gap is provided for the 10,000-year return period. This will be expected as the minimum standard for all new installations.

Offshore Information Note No.1/2008: The Role of an Offshore Metocean Advisor (2008)

This information sheet gives an example of good practice in how duty holders (operators) can deal with the wide range of metocean activities and requirements through the setting up of an appropriately informed role within their company structure (either internal or external), referred to in this guidance as a 'metocean advisor' or 'metocean focal point'.

Appendix 2
Supplemental metocean terminology

1. Winds

Wind speed scales

There are several different wind scales in use globally. These provide information on speeds, descriptions and impacts. Two such examples are the Beaufort scale and the Saffir–Simpson hurricane wind scale.

Beaufort scale

The Beaufort scale is an empirical measure for describing wind intensity and associated sea conditions and is widely used by mariners (see Figure A2.1).

BEAUFORT WIND SCALE WITH CORRESPONDING SEA STATE CODES					
Beaufort Number	Wind Velocity (Knots)	Wind Description	Sea State Description	Sea State: Term and Height of Waves (Feet)	Sea State: Condition Number
0	Less than 1	Calm	Sea surface smooth and mirror-like	Calm, glassy 0	0
1	1-3	Light Air	Scaly ripples, no foam crests		
2	4-6	Light Breeze	Small wavelets, crests glassy, no breaking	Calm, rippled 0 – 0.3	1
3	7-10	Gentle Breeze	Large wavelets, crests begin to break, scattered whitecaps	Smooth, wavelets 0.3-1	2
4	11-16	Moderate Breeze	Small waves, becoming longer, numerous whitecaps	Slight 1-4	3
5	17-21	Fresh Breeze	Moderate waves, taking longer form, many whitecaps, some spray	Moderate 4-8	4
6	22-27	Strong Breeze	Larger waves, whitecaps common, more spray	Rough 8-13	5
7	28-33	Near Gale	Sea heaps up, white foam streaks off breakers	Very rough 13-20	6
8	34-40	Gale	Moderately high, waves of greater length, edges of crests begin to break into spindrift, foam blown in streaks		
9	41-47	Strong Gale	High waves, sea begins to roll, dense streaks of foam, spray may reduce visibility		
10	48-55	Storm	Very high waves, with overhanging crests, sea white with densely blown foam, heavy rolling, lowered visibility	High 20-30	7
11	56-63	Violent Storm	Exceptionally high waves, foam patches cover sea, visibility more reduced	Very high 30-45	8
12	64 and over	Hurricane	Air filled with foam, sea completely white with driving spray, visibility greatly reduced	Phenomenal 45 and over	9

Figure A2.1 Beaufort wind scale.

Table A2.1 Saffir–Simpson hurricane scale.

Category	Wind speed (mph)	Pressure (mb)	Storm surge (ft)	Damage
1	74–95	>979	4–5	Minimal
2	96–110	979–965	6–8	Moderate
3	111–130	964–945	9–12	Extensive
4	131–155	944–920	13–18	Extreme
5	>155	<920	>18	Catastrophic

Saffir–Simpson hurricane wind

The Saffir–Simpson hurricane scale (Table A2.2) rates hurricanes on a scale from 1 to 5 based on the hurricane's sustained wind speed with associated values for pressure and storm surge. This scale also gives a descriptive estimate of property damage. Category 3 hurricanes and higher are considered major hurricanes because of their potential for significant loss of life and the level of damage.

Note that most presentations of hurricane categories give information in imperial units, i.e. mph and feet.

In the western Pacific the term typhoon is used instead of hurricane. When wind speeds are in excess of 150 mph it is known as a super-typhoon.

Polar low

A polar low is a depression that forms in polar air, often near a boundary between ice and sea. It is typically a small event, compared to the more usual extra-tropical depression, and therefore often difficult to forecast. It can generate very strong winds and rough seas.

(Line) squalls

A squall is a strong wind event, relatively common in tropical areas – for example, offshore Nigeria – and is characterised by:

- linear form
- sudden onset
- a duration of minutes and
- a rather sudden decrease in speed

A squall is often accompanied by a change in wind direction, a drop in air temperature and by heavy precipitation.

2. Waves

Internal waves

Internal waves form at the boundary between water layers of different densities (pycnocline), and usually occur in the deeper waters of the open ocean. This boundary typically separates colder, more saline water at depth from warmer, less saline water above.

Internal waves can also occur in areas where the surface is significantly less buoyant – for instance, where snow-melt leads to a very thin surface layer of cold, less dense water, or where major rivers flow into the ocean.

The seabed topography can also contribute to internal wave formation, particularly where seamounts occur in highly stratified waters. This can generate solitary wave packets (known as solitons) associated with local density variations, causing localised short-term increases in currents.

As with surface waves, internal waves also possess troughs, crests, wavelength and period. They normally propagate forward horizontally along the boundary.

The height of internal waves can be greater than 30 m, causing the pycnocline to undulate slowly through a considerable depth. Their wavelength often exceeds 0.5 km and their periods are typically 5–8 minutes.

Although internal waves are classified as non-surface features, evidence of such activity can often be seen at the surface (from satellite imagery) as near parallel lines of alternating calm and slightly rougher conditions. This is a consequence of the crests and troughs associated with the internal waves. The satellite image in Figure A2.2 illustrates the surface manifestation of internal waves in the Andaman Sea.

Tsunami

A tsunami is a series of waves created when a body of water, such as an ocean, is rapidly displaced due to, inter alia, earthquakes, mass movements above or below water, volcanic eruptions and landslides.

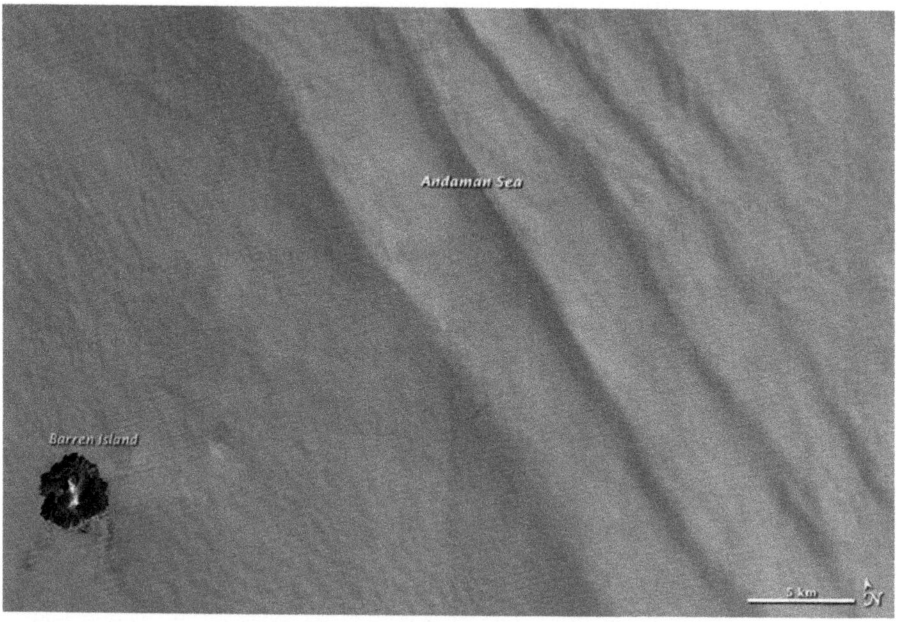

Figure A2.2 Surface view of internal waves in the Andaman Sea (courtesy of Earth Observatory, NASA).

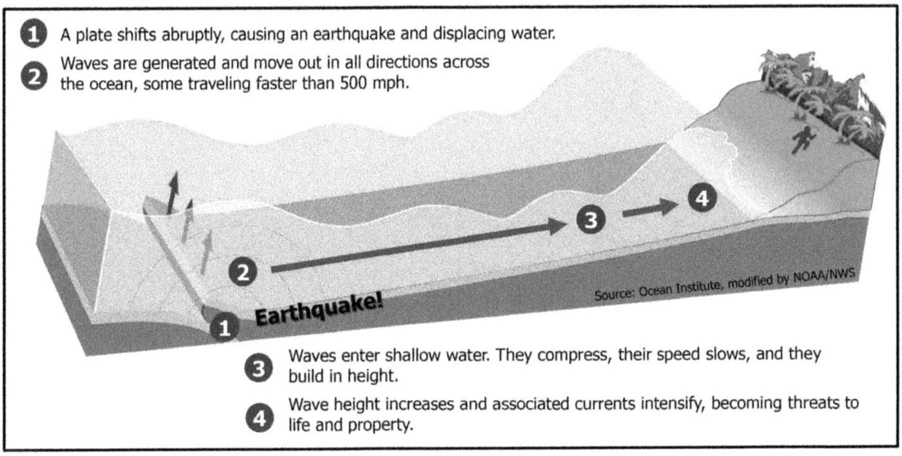

Figure A2.3 Tsunami schematic (courtesy of the Ocean Institute, modified by NOAA/NWS).

A tsunami has a much smaller amplitude offshore, with wave heights <1 m, but a very long wavelength, typically between 30 km and 300 km. Consequently, the waves generally pass unnoticed at sea, forming only a passing 'hump' in the ocean. As they approach the coast and shallow water, however, a dramatic increase in height occurs as the waves slow down and bunch up resulting in potentially catastrophic breaking waves that can be in excess of 10 m in height (Figure A2.3).

3. Currents and water levels

Forcing mechanisms: astronomical

Astronomical bodies, principally the Sun and the Moon, cause a periodic and predictable rise and fall of sea level (tide) with an associated horizontal movement of water (tidal current).

In the open ocean, where the direction of flow is unrestricted, the tidal current is rotary: that is, it flows continuously and the direction rotates through all points of the compass over the complete tidal period. This is caused by the Earth's rotation and, unless modified by local conditions (such as coasts and water depth), is clockwise in the northern hemisphere.

In semi-enclosed or narrow sea areas – for example, the North Sea – the tidal current is essentially restricted to motion in two main opposing directions, and is referred to as rectilinear. When the tide turns from one to the other, a period of reduced flow is experienced. This is termed slack water, and can be important for carrying out diving operations.

Forcing mechanisms: meteorological

The primary meteorological forcing mechanism is wind 'pushing' the sea and creating a surge (also referred to as a storm surge), which results in both an elevation effect and a horizontal current effect.

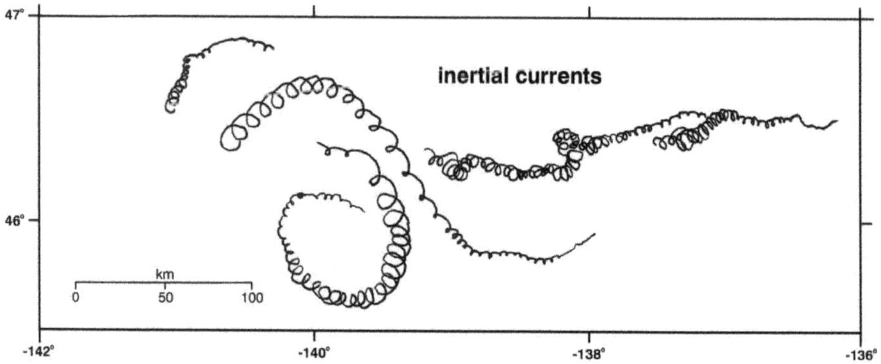

Figure A2.4 Progressive vector plots showing circular flow of inertial currents.

The storms that generate such surges are also often associated with low atmospheric pressure systems. This secondary factor can be extremely significant in some circumstances as low pressure has the effect of elevating the sea surface still further, which can lead to enhanced surges.

The response of the ocean to wind stress is not intuitive. Motions along the direction of forcing are significantly altered by the rotation of the Earth. In the northern hemisphere the Coriolis force (due to the Earth's rotation) acts to the right of the current flow. As time elapses, with decreasing winds and in the absence of other forces, this will result in clockwise circular motions (Figure A2.4) in the northern hemisphere (anticlockwise in the southern hemisphere) referred to as inertial currents, with a periodicity linked to latitude.

The inertial period equation is:

$$f = 2\Omega \sin\Phi$$

where

f = inertial frequency

Ω = angular speed of the Earth's rotation about its axis ($= 7.29 \times 10^{-5}\ s^{-1}$)

Φ = latitude

At the Equator $f = 0$. Thus, for example, the periodicity of an inertial current would be 24 hrs at 30° N and 14 hrs at 60° N.

Frictional effects significantly alter the inertial oscillations described above. For example, frictional damping reduces the amplitude of the oscillations, allowing the parcel of water to take up a component of motion closer to the forcing wind direction. The net, long-term modified surface flow (Ekman drift) is typically from 10° to 45° to the right (northern hemisphere) of the wind stress, and its magnitude can be approximated as 1–3% of the wind speed.

Forcing mechanisms: oceanographic

The primary cause of oceanographic forcing are variations in density brought about by differences in temperature and salinity. These variations create horizontal pressure gradients that, in turn, result in current flow (also known as thermohaline flow).

The strongest density gradients occur at oceanic fronts in the deeper waters of the world's oceans, and therefore such motions are often less significant in the shallower waters of shelf seas.

The resulting density-driven flows take on many forms, such as large-scale ocean circulation and (continental) slope currents, and although the mechanisms are relatively well known they can be very difficult to predict on the spatial and temporal scales required for most offshore engineering purposes.

Oceanographically-forced currents are also affected by the Earth's rotation. Currents that are a result of density variations and are modified by the Earth's rotation are termed geostrophic currents. Depending on location, the geostrophic flows comprise a depth-invariant (barotropic) component and a depth-varying (baroclinic) component.

Density-driven flows: eddies

Eddies are relatively small, contained pockets of moving water that break off from the main body of a current and travel independently of their parent. They range in diameter from a few centimetres to hundreds of kilometres.

The smallest-scale eddies might last for only a matter of seconds, while the larger can persist for days, weeks or months. The latter are more commonly known as mesoscale eddies and can form in almost any part of a current, but are especially pronounced in oceanic 'western boundary currents', such as the Gulf Stream (Figure A2.5).

Eddies are characterised by currents that flow in a roughly circular motion around the centre of the eddy. The direction of rotation of these currents may either be cyclonic (counter-clockwise) or anti-cyclonic (clockwise).

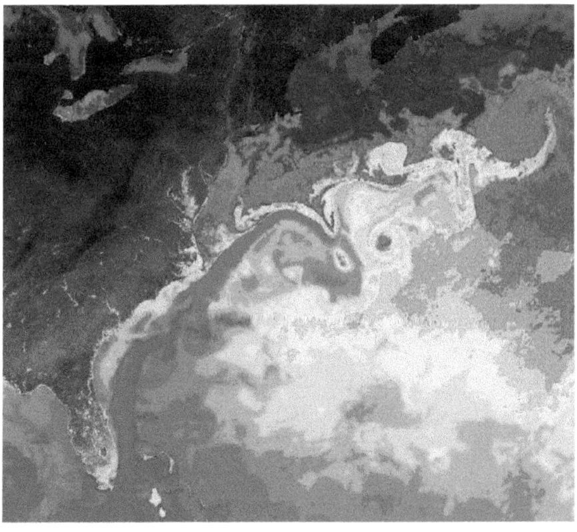

Figure A2.5 Warm-core (red/orange circle) and cold-core (blue/green) eddies in the Gulf Stream (NASA, public domain via Wikimedia Commons).

In the northern hemisphere, cyclonic rotation is associated with cold-core eddies (essentially masses of cold water surrounded by warmer water), whereas anti-cyclonic rotation is associated with warm-core eddies (masses of warm water surrounded by colder water).

Thus, an oceanic eddy is also usually made of water masses that are different from those outside it. This means that the water within an eddy usually has different temperature and salinity characteristics to the water outside it.

Because eddies may have a vigorous circulation associated with them, they are of concern to operations at sea. Current speeds in excess of 2 m/s are not uncommon.

Figure A2.6 gives an example of eddies located in the Faroes–Shetland Channel and also shows the northward movement of the Gulf Stream (red and yellow colours).

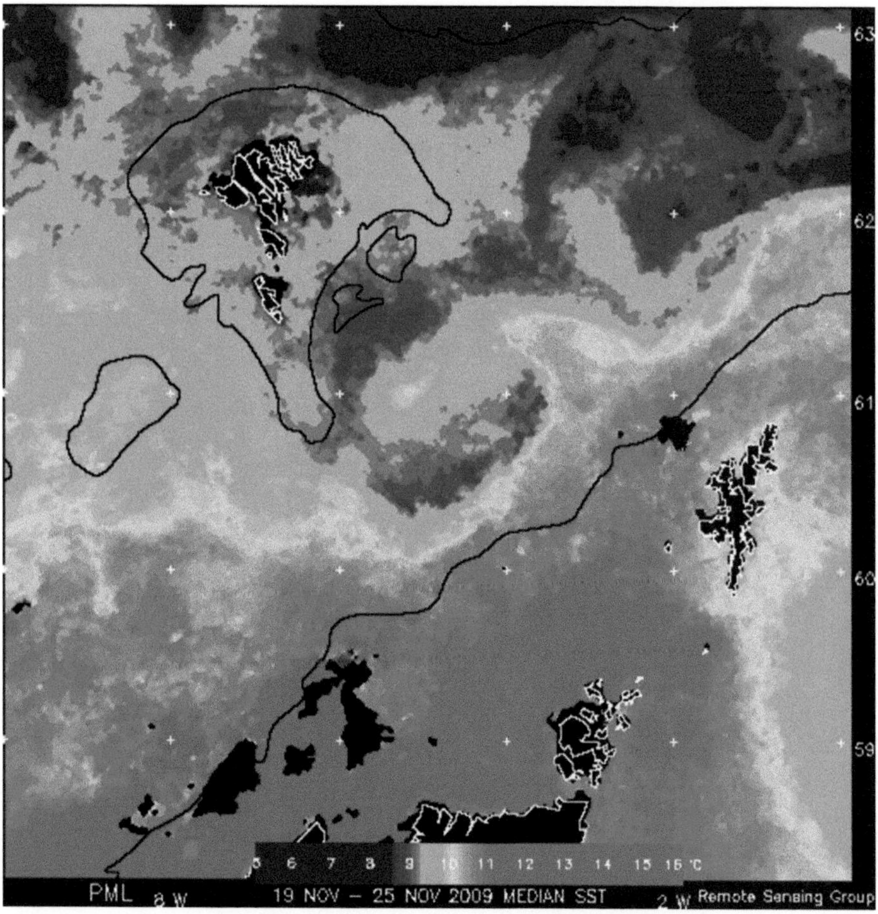

Figure A2.6 Eddies in the Faroes–Shetlands Channel (image produced by the NERC Earth Observation Data Acquisition and Analysis Service (NEODAAS) from Advanced Very-High-Resolution Radiometer (AVHRR) data).

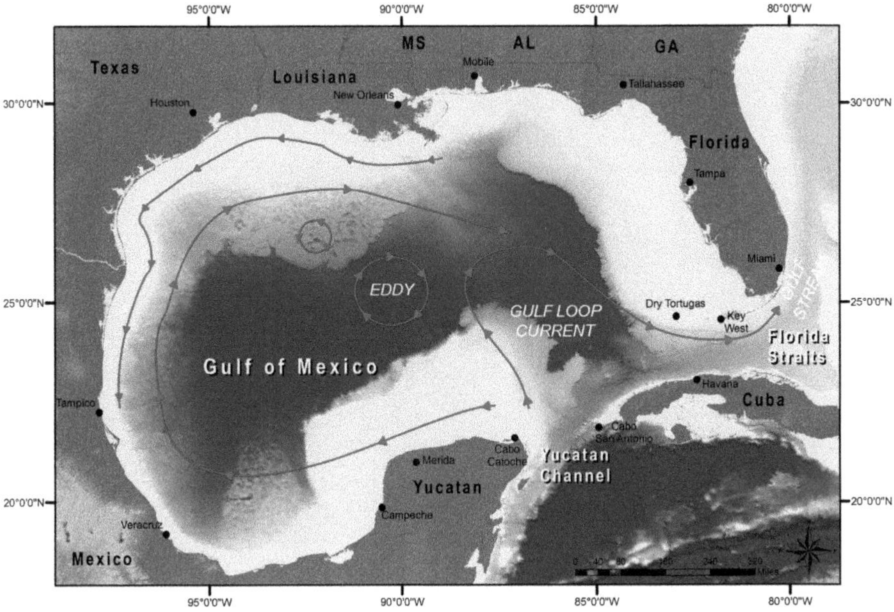

Figure A2.7 Schematic view of the Loop Current and a loop eddy (image courtesy of NOAA Flower Garden Banks National Marine Sanctuary).

The Loop Current (Gulf of Mexico) and loop eddies

The Loop Current is a warm ocean current that flows northward between Cuba and the Yucatán Peninsula into the Gulf of Mexico, which 'loops' east before exiting through the Florida straits and joining the Gulf Stream (Figure A2.7). It serves as the dominant circulation feature in the eastern Gulf of Mexico, and can reach maximum flow speeds of almost 2 m/s.

A related feature is an area of warm water that separates from the Loop Current, somewhat randomly every 3–17 months. Such phenomena are known as loop eddies, with current speeds of similar magnitude to the loop current. They can measure 200–400 kms in diameter and extend down to a depth of 1,000 m. They drift westwards at speeds of 2–5 km/day and have a lifespan of up to a year before 'hitting' the coast of Texas or Mexico.

Satellite imagery is routinely used to show these features. Figure A2.8 shows sea temperature in the Gulf of Mexico, with red being the warmest and blue the coldest. It clearly shows the loop current moving north in the Yucatan Channel and an eddy in the process of 'pinching' itself off prior to moving westwards.

Satellite imagery can also show similar details by focusing on sea surface height. Figure A2.9 shows sea surface height differences in the Gulf of Mexico. The intense eddy (red) in the centre is typically up to 30 cm higher than the average height.

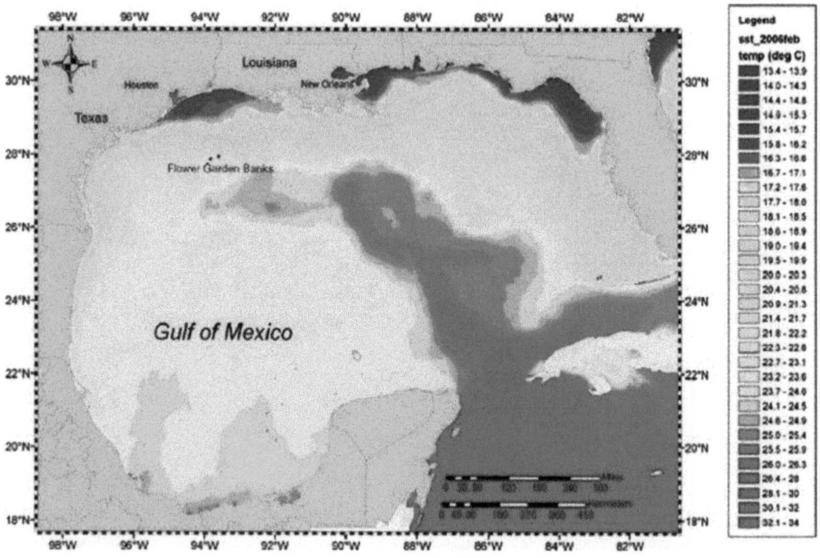

Figure A2.8 Surface sea temperatures showing features in the Gulf of Mexico (courtesy of NOAA, in Schmahl and Hickerson in *Coral Reefs of the USA*, 2008).

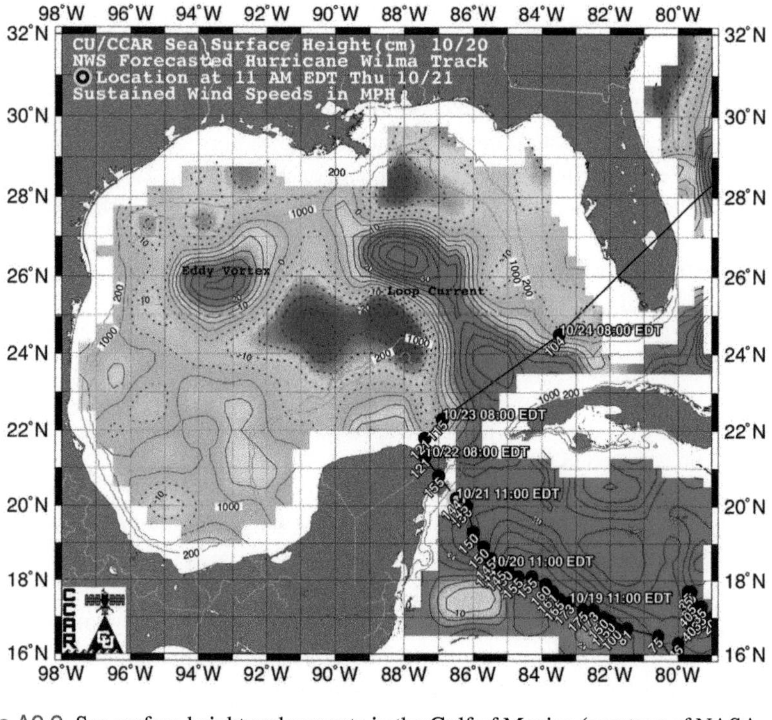

Figure A2.9 Sea surface height and currents in the Gulf of Mexico (courtesy of NASA https://sealevel.jpl.nasa.gov/files/archive/activities/ts2siac2.pdf).

Figure A2.10 A river plume – Elwha River, Washington (courtesy of USGS).

Density-driven flows: river plumes

Where a river flows out into the sea or ocean the fresh water can often be observed extending several hundred metres or even a few kilometres from the coast, particularly if the river contains high levels of sediment (Figure A2.10). This river extension is referred to as a river plume. Such plumes generally occupy a wide but shallow sea surface layer bounded by a sharp density gradient.

These significant differences in temperature, salinity and density between the plume and the adjacent open water can result in potentially complex and fast-moving currents.

Turbidity currents

Turbidity currents are rapid, downhill flows of water caused by increased density due to high amounts of sediment. They can arise when mud and sand on the continental shelf and/or slope are loosened by earthquakes and other geological disturbances. The turbid water then rushes downward like an avalanche, picking up sediment and increasing in speed as it flows. Speeds can be in excess of 20 m/sec.

Turbidity currents can change the physical shape of the seafloor by eroding large areas and creating underwater canyons. They also deposit huge amounts of sediment wherever they flow, usually in a gradient or fan pattern, with the largest particles at the bottom and the smallest ones on top.

Such currents are difficult to predict and have the potential to cause catastrophic damage to offshore installations and pipelines.

4. Nearshore/coastal terminology

In nearshore or coastal waters, various factors – such as shallowing water depths and coastline features – result in modifications to metocean conditions emanating from further offshore. This applies particularly to wave conditions, where the following phenomena can occur:

- shoaling
- breaking
- refraction
- diffraction
- wave–current interaction

Other phenomena associated with shallow-water regions can include:

- rip currents
- land/sea breezes

What is meant by shallow water?

A general rule of thumb is that if the water depth is < half the wavelength, then a wave is considered to be in shallow water. However, for most purposes shallow water tends to be applied to water depths generally less than 20–30 m.

When waves travel into areas of shallow water they begin to be affected by the seabed – processes such as shoaling and refraction occur, culminating in waves breaking when they reach the shore. Breaking waves can take several forms depending on the location and seabed slope. The 2 most common are spilling and plunging breakers. Spilling breakers are associated with a gradual seabed slope and result in gentle breaking waves, whereas plunging breakers occur with steep slopes and result in a more violent impact often characterised by a 'tube-like structure' to the breaker (see Figure A2.11).

Modification to waves

Shoaling

Shoaling is the effect by which surface waves increase in height upon entering shallower water. The free orbital motion of the water is disrupted (Figure A2.12) and water particles in orbital motion no longer return to their original position (they 'feel' the seabed). The wave slows down, the wavelength decreases, the wave period remains the same and the waves start to bunch up and steepen.

As a wave moves into increasingly shallow water there comes a point when the retardation near the seabed is overtaken by the forward motion at the surface – the wave then starts to break. This usually happens when the height of the wave becomes about the same size as the local water depth. A rule-of-thumb criterion for a breaking wave is 0.8 × water depth.

Figure A2.11 The 'tube' effect of a breaking wave (Photo 26236983 © Craig Levers | Dreamstime.com).

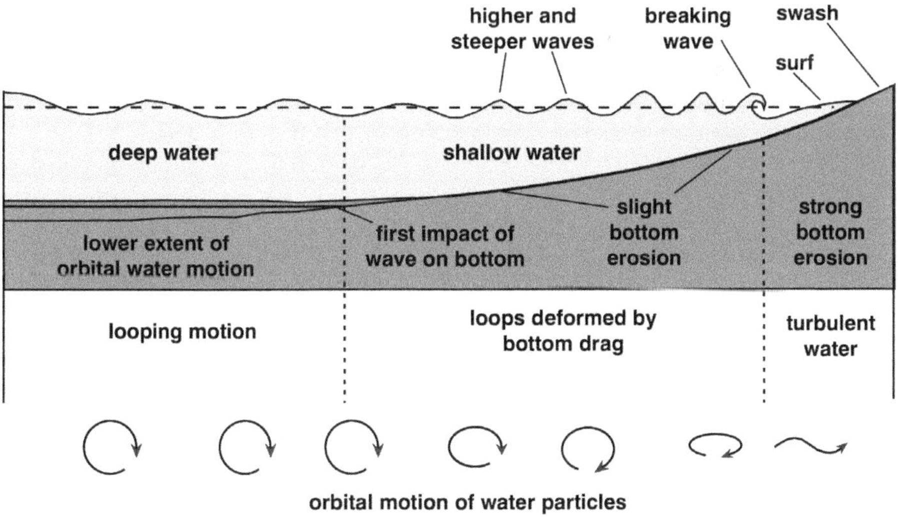

Figure A2.12 Schematic showing modification to waves approaching the shore.

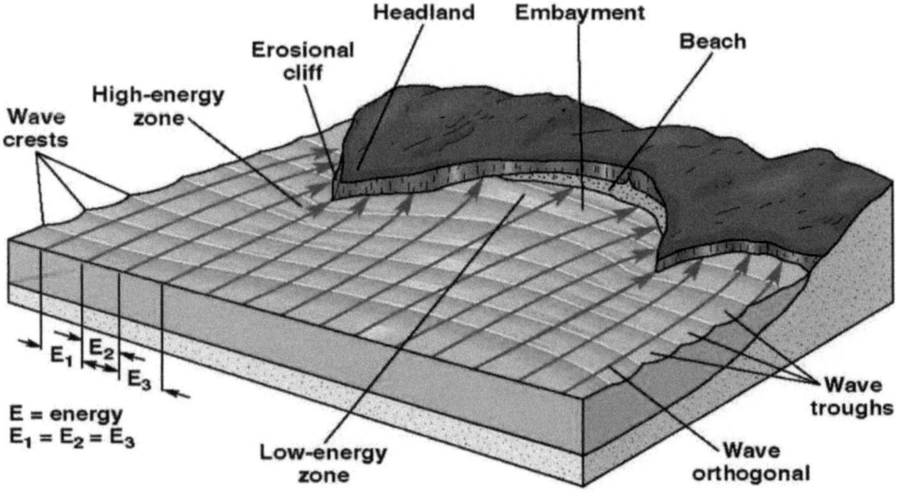

Figure A2.13 Schematic of wave refraction.

Refraction

Refraction is the process by which a wave is 'bent' or turned from its original direction as it approaches the coast at an angle. The part of the wave that first touches the seabed slows down, whereas the rest of the wave in slightly deeper water travels at its original speed – this causes the wave to bend to meet the shoreline (Figure A2.13).

Refraction causes waves to converge at headlands, where the increased energy causes erosion. Conversely, divergence and deposition of sediment occur in coves and bays.

Diffraction

Diffraction is another term concerned with waves changing direction. It can happen when waves encounter obstacles, such as a breakwater or an island (Figure A2.14). After passing the obstacle, the waves 'turn' into the region behind the obstacle and carry wave energy and the wave crest into this so-called shadow zone.

It should be noted that diffraction is not necessarily just a shallow-water phenomenon: it can happen further offshore in association with the legs of platforms and rigs.

Wave–current interaction

As its name implies, wave–current interaction is a term applied to the interaction between surface waves and a mean (current) flow. It is primarily an issue in shallow-water areas that experience strong currents or tidal flows but it can also occur in the open ocean.

The interaction implies an exchange of energy so that, once the interaction has begun, both the waves and the mean flow are affected. Thus, when a wave system encounters an area of strong currents the wave height, the wavelength, the direction and the wave energy can be altered. Likewise the current speed and direction can also change (Figure A2.15).

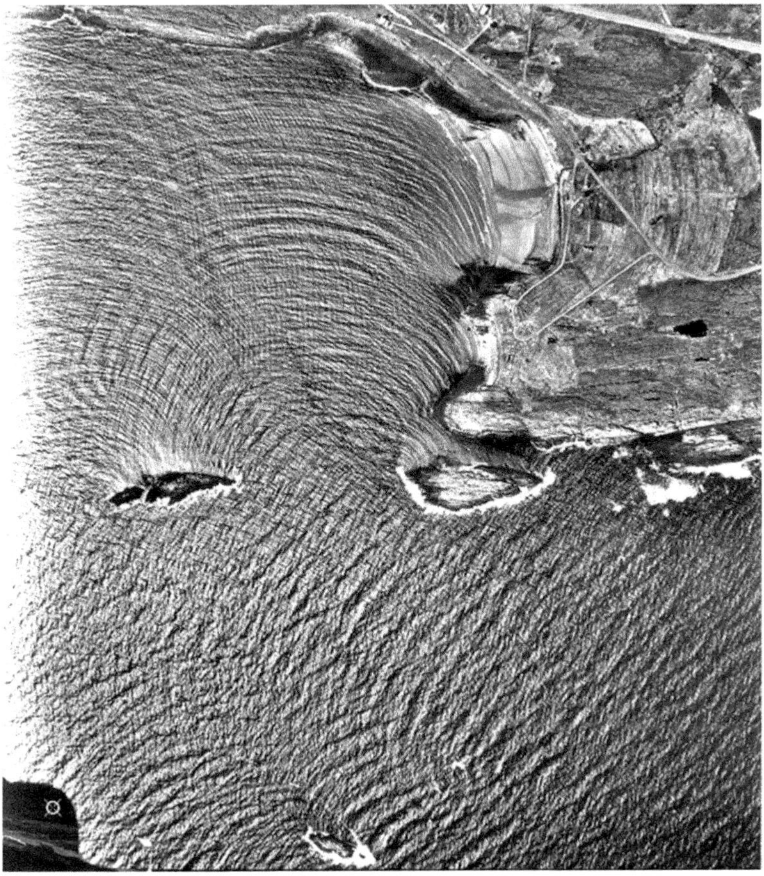

Figure A2.14 Wave diffraction caused by islands (courtesy of R. Suddeth).

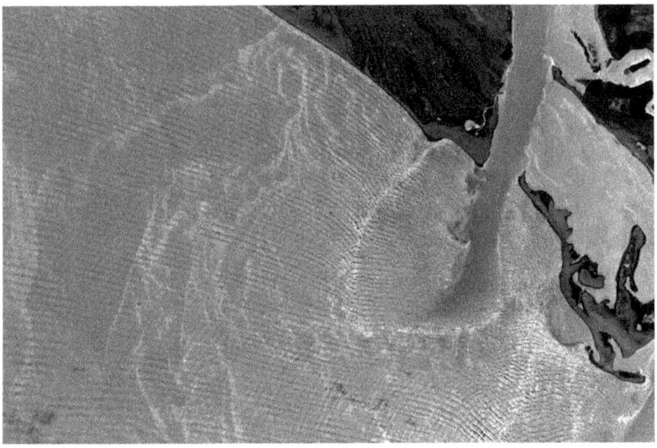

Figure A2.15 Wave sets and tidal currents, Gulf of California (courtesy of NASA).

The degree of alteration/change depends on a number of factors but the most important are the directions of each, the speed of flow and the height of waves. For example, when directions are against each other it can cause wave heights to increase, waves to steepen and cause breaking.

5. Other nearshore/coastal effects

Land/sea breezes

Land and sea breezes describe the winds that blow onshore from sea to land and vice versa at different times of the day and night (Figure A2.16). They are more common in the tropics but still occur in temperate latitudes, particularly during summer.

A sea breeze or onshore breeze is a gentle wind blowing from the sea toward the land. It develops over bodies of water near land due to differences in air pressure created by their different temperatures. It is a common occurrence along

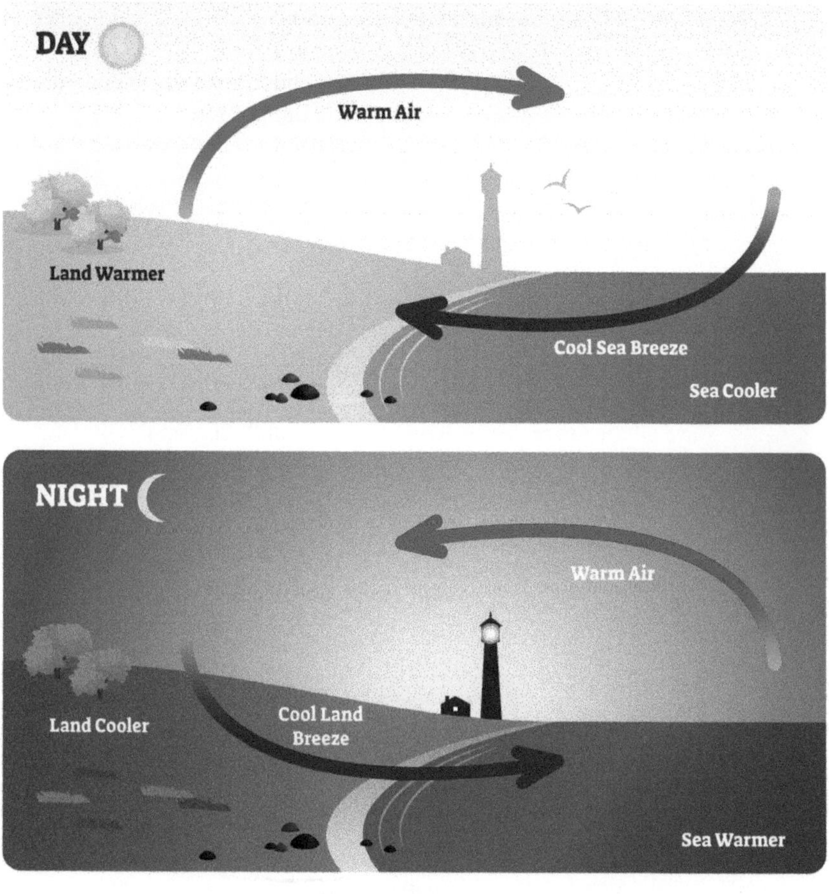

Figure A2.16 Schematic of land/sea breeze (illustration 180283643 © VectorMine | Dreamstime.com).

coasts during the morning, as solar radiation heats the land more quickly than it does the water.

A land breeze or offshore breeze is the reverse effect, blowing from the land to the sea. It is caused by the land cooling more quickly than the water in the evening. The sea breeze dissipates and the wind flows from the land towards the sea. It tends to be weaker than the sea breeze.

Rip currents

A rip current is a strong flow of water moving away from the shore and can occur at any beach with breaking waves.

When winds and waves push water towards the shore, it is often forced sideways by the oncoming waves. This water streams along the shoreline until it finds an exit back to the sea. The resulting rip current is usually narrow and located in a trench between sandbars, under piers or along jetties.

Figure A2.17 shows some of the main features associated with rip currents.

Rip currents can vary in size but typically are 50–500 m long and 5–10 m wide. Speeds are usually in excess of 0.5 m/s but can be up to 2.5 m/s. They are stronger when the surf is rough, for example, during high onshore winds, or when the tide level is low.

It is important to realise that rip currents are essentially surface phenomena. This strong surface flow tends to dampen incoming waves, leading to the illusion of a particularly calm part of the sea.

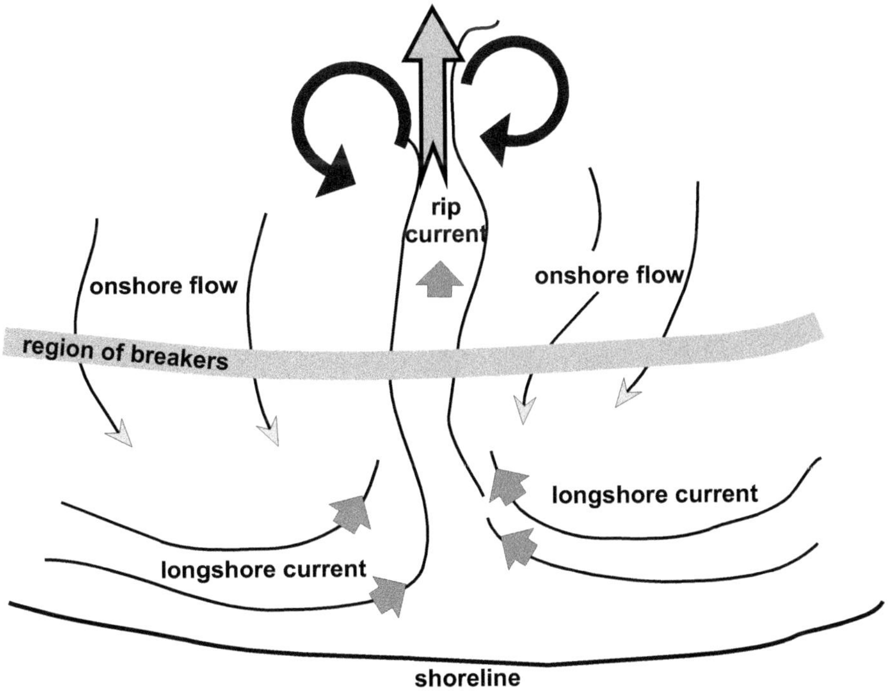

Figure A2.17 Diagram showing rip currents.

Appendix 3
Wave spectra

If we look out to sea, the waves on the sea surface are not of a simple sinusoidal form. Rather, the surface appears to be composed of random waves of various lengths and periods, sometimes criss-crossing one another. This surface can best be described via the concept of the 'spectrum of ocean waves'. The spectrum essentially defines the distribution of wave energy among different wave frequencies (equivalent to wave periods).

There are basically two different categories of wave spectra:

- uni-modal wave types, characterised by a single peak in energy (Pierson–Moskowitz spectrum, JONSWAP spectrum)
- bi-modal formulations, characterised by two peaks representing a wind–sea component and a swell component (Ochi–Hubble wave spectrum, Ochi–Hubble wave spectrum)

1. Uni-modal wave spectra

The Pierson–Moskowitz (PM) spectrum is an empirical relationship that defines the distribution of energy with frequency within the ocean. It was developed in 1964 from measurements made in the North Atlantic and is one of the simplest descriptions for the energy distribution.

It assumes that, if the wind blows steadily for a long time over a large area, then the waves will eventually reach a point of equilibrium with the wind. This is known as a fully developed sea.

However, Hasselmann *et al.* (1973), after analysing data collected during the Joint North Sea Wave Observation Project (JONSWAP), found that the wave spectrum is never fully developed. Rather, it continues to develop through non-linear, wave–wave interactions even for very long times and distances. Hence an extra and somewhat artificial factor was added to the Pierson–Moskowitz spectrum in order to improve the fit to their measurements.

The JONSWAP spectrum is thus effectively a fetch-limited version of the Pierson–Moskowitz spectrum multiplied by an extra peak enhancement factor (γ). The default value for γ is 3.3 based on the mean values from the data of the original JONSWAP project in relatively deep water (50–100 m).

When fitting measured data to the JONSWAP spectrum, the values derived for γ vary widely between different times during the development of the sea and between different sites around the world. However, a central value of $\gamma \approx 2$ is generally appropriate for very severe storms.

Figure A3.1 gives a comparison of (generalised) typical shapes for Pierson–Moskowitz and JONSWAP spectra, showing the enhanced peak of the latter.

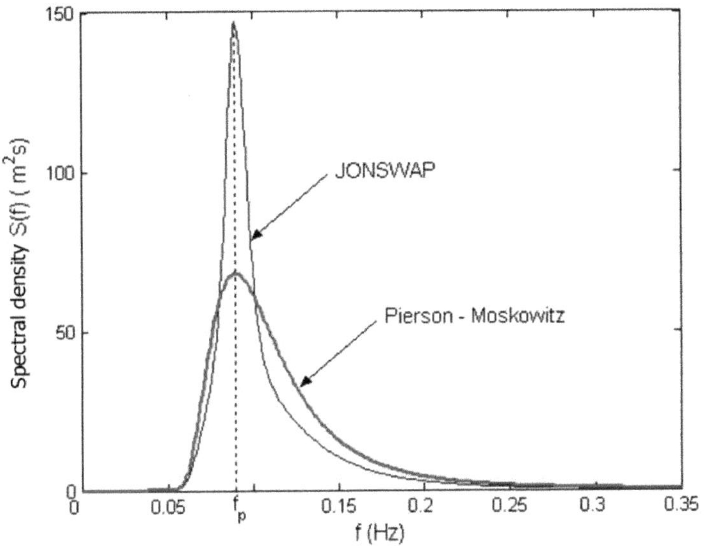

Figure A3.1 Comparison of Pierson–Moskowitz and JONSWAP spectra.

Guidance for the use of an appropriate wave spectrum is contained in *ISO 19901-1*:

> The most appropriate form of the wave frequency spectrum for an offshore structure depends on the geographical area, the severity of the sea state, whether the sea state is fully developed or is still growing, and the application concerned.
>
> For example, for a short-term North Sea design storm condition, a unidirectional JONSWAP spectrum can be most appropriate, whereas for the modelling of a series of sea states for a long-term fatigue analysis directionally spread Pierson–Moskowitz spectra are often more appropriate.

2. Bi-modal wave spectra

The previous models cannot fully represent multi-peaked spectra. The simplest way to represent such spectra is to decompose the spectrum into two parts: a wind–sea component and a swell component (Figure A3.2).

The Ochi–Hubble wave spectrum

The Ochi–Hubble wave spectrum (Figure A3.3) is the sum of two separate component spectra each of which is described by a further generalised (3-parameter) Pierson–Moskowitz spectrum. The two component spectra are therefore each specified by a set of 3 parameters: H_{s1}, ω_{m1}, λ_1 for the lower-frequency component; and H_{s2}, ω_{m2}, λ_2 for the higher-frequency component. Here

- H_{s1} and H_{s2} are the significant wave heights of the component spectra; the overall significant wave height $H_s = (H_{s1}^2 + H_{s2}^2)^{½}$

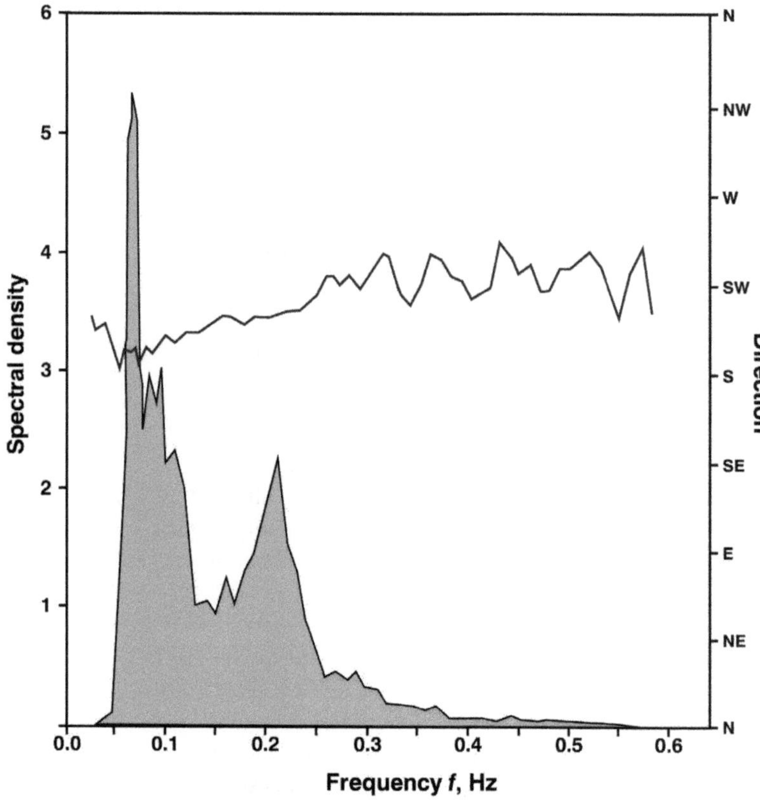

Figure A3.2 An example of a measured double-peaked (bi-modal) spectrum. The highest peak represents the swell component, the lower peak the wind–sea component, and the blue line the wave direction.

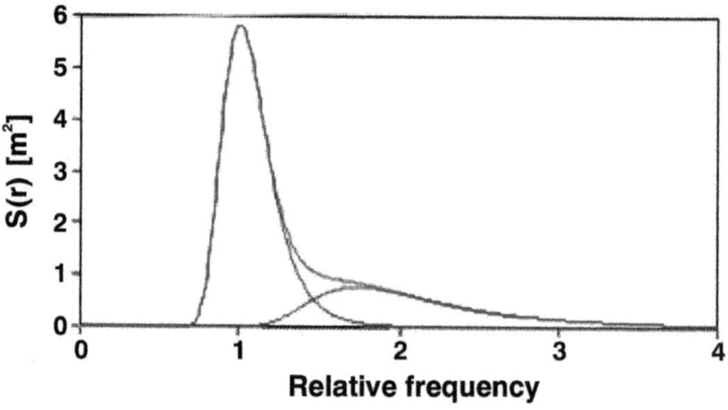

Figure A3.3 An example of the Ochi–Hubble spectrum.

- ω_{m1} and ω_{m2} are the modal angular frequencies of the two components
- λ_1 and λ_2 are shape parameters that control the extent to which the spectral energy is concentrated around the modal frequency – larger values give more concentrated component spectra

The Torsethaugen spectrum

The Torsethaugen spectrum is a double-peak spectral model based on measured spectra for Norwegian waters and is therefore more suited to North Sea application than Ochi–Hubble. It is also based on a combination of two single-peak spectra describing a wind–sea and a swell–sea system. Each sea system is defined by five parameters: Hs, Tp, γ, M and N, all of which are parameterised by Hs and Tp.

3. The occurrence of uni-modal vs. bi-modal spectra

In most sea areas the proportion of sea states tending to bi-modality is of the order of 20%. For example, Soares (1984, 1991) indicated values of 16% for the North Sea. However, this percentage typically varies with wave height and usually decreases with increasing height (Massel, 1996). During storm conditions such spectra are typically found in less than 10% of sea states.

More recent analysis (post-2000) of sea states from the North Sea for both the NEXTRA and NORA10 hindcasts suggest that uni-modal wave spectra are perfectly adequate at describing extreme sea states. There is, typically, convergence of the ratio of Tp (wind-sea or swell) to Tp (whole sea) towards unity at high sea states. As the whole sea, wind–sea and swell peak periods are converging to similar values the wave spectra are by definition tending to uni-modality. This suggests that extreme sea states are well represented by a uni-modal JONSWAP wave spectrum with $\gamma \approx 2.2$.

4. Deriving engineering parameters from spectra

Spectra can be further condensed into 'spectral moments', from which a number of parameters can be derived. Although spectra are usually plotted as a line, the more correct way of plotting them would really be as a histogram.

The spectral moments are derived from spectra as follows:

(i) determine the frequency that each column in the spectral histogram represents
(ii) raise this frequency to a selected power: -2, -1, 0, 1, 2, 3, 4
(iii) multiply the area of the column by the result of (ii) above

The resulting spectral moments are referred to as: m-2, m-1, m0, m1, m2, m3 and m4.

The zero moment (m0) is the easiest to understand, as any number to the power of zero is 1. Therefore, in every case, the frequency of the individual columns raised to the power zero is 1. Thus, the areas of the columns are multiplied by 1 and summed, and the result is simply the area under the curve. In fact, the zero moment is probably the most important, as $4\sqrt{m0}$ is the significant wave height (Hs), usually abbreviated as H4rm0 (or Hm0) when it is derived from spectra.

The second moment (m2) has all of the frequencies squared before being multiplied by the column heights and summed. This moment is used in the derivation of mean wave period (Tz), which is $\sqrt{(m2/m0)}$.

Other parameters may be derived by various combinations of spectral moments. However, one must always bear in mind that the higher-order moments are very much more sensitive to small errors in the spectrum, as the frequencies might, for example, be raised to the fourth power before being multiplied by the area of the column.

In fact, both Hs and Tz could be derived deterministically from the time series. However, there is another period parameter that is unique to spectra: the spectral peak period, Tp. This is simply the period that is associated with the peak of the spectrum.

Sometimes more than one Tp might be derived and different companies have, in the past, used the same abbreviation to mean different things. Therefore, if two estimates of Tp appear in the data listing, great care must be taken to establish exactly what each is.

For example, Tp1 might be the highest energy peak *or* the highest frequency peak. Interestingly, the ratio between Tp and Tz is close to the ratio between Tass (the period associated with the highest individual wave in the sample) and Tz.

References

Hasselmann K., Barnett, T.P., Bouws, E., Carlson, H., Cartwright, D.E., Enke, K., Ewing, J.A., Gienapp, H., Hasselmann, D.E., Kruseman, P., Meerburg, A., Muller, P., Olbers, D.J., Richter, K., Sell, W. and Walden, H. (1973) Measurements of wind-wave growth and swell decay during the Joint North Sea Wave Project (JONSWAP). *Ergnzungsheft zur Deutschen Hydrographischen Zeitschrift* Reihe A(8) (Nr. 12), p. 95.

Massel, S.R. (1996) *Ocean Surface Waves: Their Physics and Prediction.* Advanced Series on Ocean Engineering, volume 11. Totnes, Devon: World Scientific Publishing.

Ochi, M.K. and Hubble, E.N. (1976) Six-parameter wave spectra. *Proc 15th Coastal Engineering Conference*, pp. 301–328.

Soares, C.G. (1984) Representation of double-peaked sea wave spectra. *Ocean Engineering*, 11, pp. 185–207.

Soares, C.G. (1991) On the occurrence of double-peaked wave spectra. *Ocean Engineering*, 18, pp. 167–171.

Torsethaugen, K. (1996) Model for a doubly peaked wave spectrum. Technical Report STF22 A96204, Trondheim, Norway: SINTEF Civil and Environmental Engineering.

Torsethaugen, K. and Haver, S. (2004) Simplified double peak spectral model for ocean waves. JSC Paper 193.

Appendix 4
Extreme value analysis

Extreme value analysis (EVA) is a technique employed to extrapolate measured or hindcast data to very low probabilities of occurrence, equivalent to n-year return periods, where n is normally between 1 and 100 years.

The basis behind EVA is approximating a probability distribution function to the distribution of (time-limited) data. This function is then extrapolated and evaluated at the required level of probability.

There are two main methods commonly used: cumulative frequency extrapolation (CFE) and peaks-over-threshold (POT).

It must be stressed that there is no physical or statistical support for the use of any of the analytical functions used. The method is entirely empirical and is based on the assumption that, if the chosen function or distribution fits the data reasonably well over the measured range, it will continue to do so at the extremes. However, it should be noted that there is a preference for POT, which tends to give good results from long-term quality controlled datasets.

1. Cumulative frequency extrapolation

Cumulative frequency extrapolation (CFE) involves grouping all the parameter values in the dataset using specified class intervals and then forming a cumulative frequency distribution (CFD) by summing the number of observations greater than or equal to the lower bound of the class interval. For example, with significant wave height (Hs) data a typical class interval would be 0.5 m.

Suitable probability distributions (see Section 3) are then fitted to the data, using either the method of least squares, method of moments or maximum likelihood, in order to extrapolate to the required probability of non-exceedance.

The advantage of the CFE method is that it can be used with as little as one year of data, as it uses all the data samples. However, the probability levels calculated for this method assume that the measurements are independent and identically distributed. Therefore, by ignoring any possible correlation between consecutive values of the measurements, it has been suggested that this method may result in an under-estimation of extreme values (Carter & Challenor, 1981).

However, despite the concerns of Carter and Challenor, it can still give good results if used with care. This method was widely used in the 20th century but has since tended to be superceded by the use of Peaks-over-Threshold analysis.

2. Peaks-over-threshold analysis

The peaks-over-threshold technique (POT) involves the production of a subset of the entire dataset by only selecting storm peak events that

- exceed a predetermined threshold (for example, Hs values >7 m), and
- are separated in time by a specified number of hours (typically 18–24)

The purpose of the time separation is to ensure that peak values belonging to different events are considered independent.

The number of peaks exceeding a given level, divided by the number of years of data, gives the rate of exceedance, which can then be used to find the expected number of occurrences in a specified length of time.

The probability distribution of the peak values is then combined with the rate of occurrence of peaks to give the unconditional distribution of peak values from which *n*-year extreme values are calculated, i.e.

$$P(x.y) = P(x/y)\,P(y)$$

where

$P(x.y)$ is the unconditional probability distribution of peak values with time,
$P(x/y)$ is the conditional probability distribution of peak values, and
$P(y)$ is the probability distribution of storms with time.

The return periods (RP) of extreme values are calculated as follows:

$$RP = 1/\{\lambda[1 - P(x/y)]\}$$

where λ is the Poisson parameter and $P(x/y)$ is the conditional probability distribution of peak values.

The number of storms occurring per unit time is assumed to be a random variable that may be represented by the Poisson distribution.

The Poisson distribution is characterised by a mean value, λ, which is the average number of storms per year. The probability density of the Poisson distribution is given by the following formula:

$$p(i) = (\lambda^{i}\,\exp^{-i})/i!$$

where $i = 0, 1, 2, \ldots$

There are several drawbacks with the POT approach:

- the definition of what constitute separate 'events' is rather arbitrary
- the results can be sensitive to the threshold selected
- the restriction to peak events means that useful information can be lost, e.g. about the relationship to other variables like the wave or wind direction
- the approach cannot deal satisfactorily with data that contain gaps or with irregularly sampled data such as data from satellites

3. The probability distributions

Functions commonly used for the estimation of extreme values are:

- Fisher–Tippett distribution, Type 1 (Carter & Challenor, 1981; Carter *et al.*, 1986; Johnson & Kotz, 1970; NERC, 1975)
    - o also known as a Gumbel, double exponential, Jenkinson Type 2, and extremal type 1

- Fisher–Tippett distribution, Type 2 (Carter & Challenor, 1981[1]; Johnson & Kotz, 1970[4]; NERC, 1975[5])
  - o also known as a Cauchy type, Frechet and Jenkinson Type 1
- Fisher–Tippett distribution, Type 3 (Carter & Challenor, 1981[1]; Carter *et al.*, 1986[3]; Johnson & Kotz, 1970[4]; NERC, 1975[5])
  - o also known as a Jenkinson Type 3
- Weibull (Carter & Challenor, 1981[1]; Carter *et al.*, 1986[3]; Johnson & Kotz, 1970[4])
- Exponential (Coastal Engineering Research Center, 1985[2])
- Generalised Pareto (Davison & Smith, 1990[vi])

Notes on the functions

The Generalised Pareto, Fisher–Tippett Type 1 and Exponential functions are 2-parameter distributions, whereas the Fisher–Tippett Types 2 and 3 are described by 3 parameters.

The Weibull function may be described by 2 or 3 parameters:

- α is the location parameter and the limiting value of the distribution
- β is the scale parameter and defines the spread of the distribution
- γ is the shape parameter and describes the asymmetry of the distribution

For the majority of metocean applications the Weibull distribution is the most commonly used as it tends to provide the best fits to data.

References

Carter, D.J.T. and Challenor, P.G. (1981) Estimating return values of wave height. IOS Report No. 116. Unpublished manuscript. Godalming, Surrey: Institute of Oceanographic Sciences

Carter, D.J.T. *et al.* (1986) *Estimating Wave Climate Parameters for Engineering Applications.* Offshore Technology Report No. OTH 86 228. London: HMSO.

Coastal Engineering Research Center, US Army Engineer Waterways Experiment Station (1985) Computer Program: WAVDIST (MACE 17) Extremal significant wave height distributions. Coastal Engineering Technical Note CETN-I-40, December 1985. Washington, DC: Coastal Engineering Research Center, US Army Engineer Waterways Experiment Station.

Davison, A.C. and Smith, R.L. (1990) Models for exceedences over high thresholds. *Journal of the Royal Statistical Society, Series B*, 52 (3), pp. 393–442.

Johnson, N.L. and Kotz, S. (1970) *Continuous Univariate Distributions – 1.* Boston, MA: Houghton Mifflin.

Inst of Hydrology (1975) *Flood Studies Report Volume 1: Hydrological Studies.* London: National Environmental Research Council.

Appendix 5
Snow and ice

This appendix provides information on data and design issues pertinent to O&G activities associated with snow and ice.

Snow and ice are generally classified into two main categories:

- sea ice and icebergs
- snow and ice accretion

It should be noted that the study of ice is complex, as it takes many forms and exhibits a range of features. This is particularly so at the moveable boundary between land and sea ice. Figure A5.1 illustrates this complexity and some of the features.

There is also another 'moveable boundary' to consider: that between the ice edge and open water. This has many implications: for example, changes in fetch, vessel access etc.

1. Sea ice and icebergs

Sea ice and icebergs can affect the design and operation of structures. They should be considered not only in polar areas but also in those areas subject to cold winters. The latter include the North Caspian, parts of Canada and Alaska, Sakhalin and eastern Russia, and parts of northern Europe.

Before commencing design for, construction of, or operations on, structures in areas that are likely to be affected by sea ice and icebergs, adequate data should be collected, including:

- the type of sea ice whether first-year, multi-year or of glacial origin
- variations in the thickness, consolidation and concentration of ice types, both seasonally and from year to year

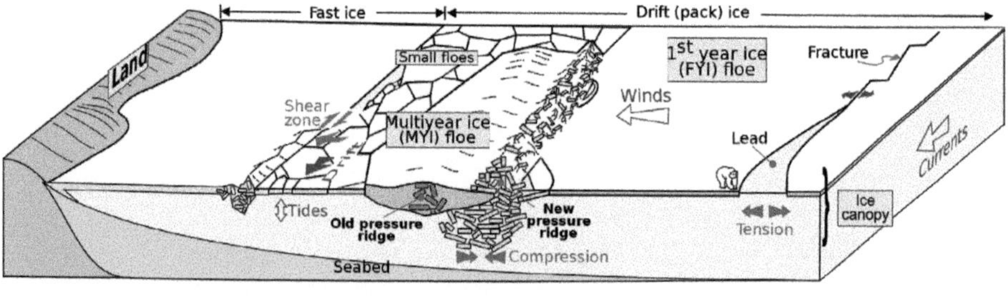

Figure A5.1 Features found in pack ice (public domain).

- variations in year-round regional ice cover, including the occurrence and distribution of ice concentrations, thicknesses, floe sizes and types present during freeze-up, winter, break-up and open-water seasons
- the probability of occurrence of specific ice features such as multi-year hummock fields and ice islands
- the distribution and probability of ice floes, pressure ridges and/or icebergs, together with their drift speed, direction, shape and mass
- in areas where ice of glacial origin is to be expected, the annual and seasonal variation in the flux, concentration and size of icebergs
- the effect of ice gouges on the seabed from icebergs or ice ridges
- the probability distributions or extreme values of the velocity of pack ice, ice floes, and discrete ice features, such as icebergs, 'bergy bits', and 'growlers', and seasonal variations of these distributions (definitions of icebergs, bergy bits and growlers relate to their size: icebergs are >5 m across, bergy bits 2–5 m and growlers are <2 m)

These data can be used to determine the design characteristics of the structure as well as possible evacuation procedures. For more specific advice on sea ice and icebergs, see *ISO 19906*.

2. Snow and ice accretion

Snow accumulation and ice accretion should be considered in the design of structures for areas that are prone to these phenomena.

Snow can settle on both horizontal surfaces and, if the snow is sufficiently wet, on non-horizontal windward parts of a structure. On vertical surfaces, it is only likely to stay in position as snow for a few hours, although it can freeze and remain as ice. It can therefore affect all exposed areas above the splash-zone. On horizontal surfaces, dry snow is blown off as soon as any thickness accumulates, while wet snow can remain in position for several hours.

An estimate should be made of the extent to which snow can accumulate on a structure and its topsides and of the possible effect on the structure.

Topsides icing can increase the diameter of structural components and lead to a substantial increase in actions caused by wind and gravity, particularly for long, slender structures such as flares. Topsides snow and icing can also adversely affect the stability of floating structures and the operation of emergency equipment.

Icing from sea spray, freezing rain or drizzle, freezing fog, or cloud droplets should also be considered in the design.

(a) Combined accumulations of snow and ice on vertical surfaces

Above deck level: The possibility of wet snow overlying a layer of ice frozen from an old snow accumulation should be considered.

Below deck level, above splash zone: The possibility of wet snow overlying a layer of ice frozen from an old snow accumulation should be considered only for the region above the splash zone.

Below deck, splash zone: Wet snow accumulations will only occur at the higher levels of the platform legs. In the splash zone the spray generated by breaking waves will wet the legs of the platform, in which circumstance the wet snow will not stick but will slide down the leg surface into the sea. At higher levels, in those regions where wet snow may adhere to the leg, the sea-spray ice accretion will be relatively small.

For more specific advice on snow and ice accretion see *ISO 19906.*

(b) Example of snow and ice accretion for the North Sea

The build-up of snow and ice is a rare event at offshore locations in the North Sea. Furthermore, measuring such an event is difficult. The lack of reliable climatological data and calibrated modelling techniques means that it is challenging to make accurate predictions of likely snow and ice accumulation on offshore structures under extreme conditions.

The recommendations and values given in Tables A5.1–A5.5 are taken from the UK HSE *Offshore Technology Report (OTR) 2001/010: Environmental Considerations* (2001), and the Norwegian Petroleum Directorate (1994).

UK HSE criteria are generally specified for three zones, but they exclude regions close to the coasts of continental Europe:

- 52° N to 54° N
- 54° N to 57.5° N
- UK sector north of 57.5° N

Norwegian Petroleum Directorate criteria apply for North Sea waters within the Norwegian sector of the North Sea only. Data are generally specified for two zones:

- 56° N to 68° N
- Norwegian sector north of 68° N

The UK HSE presents values that have been predicted from a model covering the North Sea waters west of 3° E. The model is fully described in *OTH 89 299*, entitled ‘Theoretical calculations based on climatological data’.

It should be noted that these guidance documents tend to focus on fixed structures. For floating structures, consideration needs to be given to an uneven distribution of ice and its possible effect on their stability.

UK HSE criteria (2001)

Table A5.1 UK waters from 52° N to 54° N.

	UK Department of Energy Guidance Notes					
	Wet snow		Ice from freezing spray		Ice from frozen snow	
North Sea general from 52° N to 54° N	Thickness (mm)*	Density (kg/m³)	Thickness (mm)*	Density (kg/m³)	Thickness (mm)*	Density (kg/m³)
Tubular member below deck level	–	–	25	850	–	–
Tubular member below deck level	40	500	–	–	30	900
Lattice member above deck level	40	500	–	–	25	900
Horizontal surface	200	100	–	–	–	–

* Note that the term thickness relates, in the case of tubular members, to the increase in the radius of the member.

Table A5.2 UK waters from 54° N to 57.5° N.

	UK Department of Energy Guidance Notes					
	Wet snow		Ice from freezing spray		Ice from frozen snow	
North Sea general from 54° N to 57.5° N	Thickness (mm)	Density (kg/m³)	Thickness (mm)	Density (kg/m³)	Thickness (mm)	Density (kg/m³)
Tubular member below deck level	–	–	15	850	–	–
Tubular member below deck level	40	500	–	–	25	900
Lattice member above deck level	40	500	–	–	10	900
Horizontal surface	240	100	–	–	–	–

* Note that the term thickness relates, in the case of tubular members, to the increase in the radius of the member.

Table A5.3 UK waters north of 57.5° N.

	UK Department of Energy Guidance Notes					
	Wet snow		Ice from freezing spray		Ice from frozen snow	
North Sea general north of 57.5° N	Thickness (mm)*	Density (kg/m³)	Thickness (mm)*	Density (kg/m³)	Thickness (mm)	Density (kg/m³)
Tubular member below deck level	–	–	25	850	–	–
Tubular member below deck level	40	500	–	–	30	900
Lattice member above deck level	40	500	–	–	25	900
Horizontal surface	200	100	–	–	–	–

* Note that the term thickness relates, in the case of tubular members, to the increase in the radius of the member.

Extreme snow and ice accumulations on offshore structures: Norwegian Petroleum Directorate (NPD) Criteria (1994)

Table A5.4 Norwegian waters from 56° N to 68° N.

Norwegian Petroleum Directorate					
	Wet snow	Ice from freezing spray		Ice from frozen rain/snow	
Norwegian waters from 56.0° N to 68.0° N	Load (kPa)	Thickness (mm)*	Density (kg/m³)	Thickness (mm)*	Density (kg/m³)
5–10 m ASL	0.5	80	850	10	900
10–25 m ASL	0.5	Linear reduction from 80 to 0	Linear reduction from 850 to 500	10	900
>25 m ASL	0.5	0	-	10	900

* Note that the term thickness relates, in the case of tubular members, to the increase in the radius of the member.

Table A5.5 Norwegian waters north of 68° N.

Norwegian Petroleum Directorate					
	Wet snow	Ice from freezing spray		Ice from frozen rain/snow	
Norwegian waters north of 68.0° N	Load (kPa)	Thickness (mm)*	Density (kg/m³)	Thickness (mm)*	Density (kg/m³)
5–10 m ASL	0.5	150	850	10	900
10–25 m ASL	0.5	Linear reduction from 150 to 0	Linear reduction from 850 to 500	10	900
>25 m ASL	0.5	0	-	10	900

* Note that the term thickness relates, in the case of tubular members, to the increase in the radius of the member.

Index